Student's C

Regents Review

Algebra 2/Trigonometry

Henry Gu

Mathematics Teacher
John Dewey High School
Brooklyn, New York

Disclaimer: The contents of this book are the author's alone and not those of the New York City Department of Education.

Author: Henry Gu
Editor: Christopher Gu

www.hsmathreview.com

Copyright © 2010 by Henry Gu. All rights reserved.

ISBN-10: 1460983874   ISBN-13: 9781460983874

"Everything should be made as simple as possible, but not simpler."

- Albert Einstein

Preface

When students have only 4 to 6 weeks to review for the Regents exams, they cannot benefit from the lengthy review books and overwhelming information from the many math websites. Our students need one review book that should be concise and efficient to help them succeed with high scores on the test. This is that book.

"Different books, different results." This book reviews all the important math topics and uses real Regents questions and shows all the necessary steps to solve these problems. Its clear format is like no other.

This book is structured in three parts:
1. A review section that will help students remember all the key topics and build their problem solving skills through the use of examples.
2. A practice section with real Regents questions.
3. Answers and explanations.

The topics for the practice questions correspond to the sections in the review section. Students can easily refer back to the matching review topics, while they are doing the practice.

I have already used these review sheets with my own Regents classes and I have seen firsthand that their performance is significantly higher than the statewide average. Both students and teachers like these review sheets because they are straightforward and practical.

For updates and information on additional titles, please visit our website www.hsmathreview.com.

Acknowledgement

Thanks to the teachers and students at John Dewey High School who have already used these review sheets for their own Regents review and have achieved excellent scores.

Thanks to my family for their unconditional love and support.

Dedication

This book is dedicated to all the students taking the Regents exams. I wish you the best of luck!

Algebra 2 and Trigonometry Review

Contents

PART 1. Algebra

I. ALGEBRAIC EXPRESSIONS, EQUATIONS, AND INEQUALITIES ..... 1.
   1. Factoring Polynomials ..... 1.
   2. Quadratic Equations, Inequalities, Quadratic-Linear Systems ..... 1.
   3. Rational Expressions, Operations, and Equations ..... 2.
   4. Radicals, Rational Exponents, and Radical Equations ..... 3.
   5. Absolute Value Equations ..... 4.
   6. Complex Numbers ..... 4.
   7. Roots of the Quadratic Equations ..... 5.
   8. Polynomial Equations of Higher Degrees ..... 5.
   9. Sequence and Series, Sigma Notation $\sum$ ..... 5.

II. RELATIONS AND FUNCTIONS ..... 7.
   1. Functions ..... 7.
      Domain and Range, Interval Notation, Function Notation,
      Vertical Line and Horizontal Line Tests, One-to-One, Onto
   2. Composition of Functions ..... 8.
   3. Inverse Functions ..... 8.
   4. Functions under a Transformation ..... 9.
   5. Important Functions and Relations ..... 9.
      (1). Direct Variation (2). Inverse Variations
      (3). Absolute Value Functions (4). Quadratic Functions and Parabolas
      (5). Equations of Circles (6). Exponential Functions and Equations
      (7). Logarithmic Functions and Equations

PART 2. Trigonometry

III. TRIGONOMETRIC FUNCTIONS ..... 12.
   1. Degrees and Radians, Coterminal Angles, Arc Length ..... 12.
   2. Trigonometric Fuctions ..... 13.
      Cofunctions, Unit Circle, Reference Angle

IV. TRIGONOMETRIC GRAPHS ..... 14.
   1. Graphs of Trigonometric Functions ..... 14.
   2. Graphs of the Reciprocal Functions ..... 14.
   3. Amplitude, Period, and Frequency ..... 15.
   4. Inverse Trigonometric Functions ..... 15.

Algebra 2 and Trigonometry Review

V. TRIGONOMETRIC APPLICATIONS — **16.**
1. Trigonometric Identities — 16.
    Two Angles, Double Angles, Half Angles
2. Trigonometric Equations — 16.
3. Applications — 17.
    Area, Law of Sines, Law of Cosines

PART 3. Probability and Statistics

VI. PROBABILITY — **18.**
1. Probability and Counting Principle — 18.
    (1)The probability of a simple event (2) Counting Principle
2. Permutation and Combination — 18.
3. Binomial Probability (Bernoulli Experiment) — 19.
    Binomial Expansions

VII. STATISTICS — **19.**
1. Statistics (Univariate Data) — 19.
    Collect Data, Mean, Median, Mode, Quartile,
    Variance, Standard Deviation, Normal Distribution, Z-Score
2. Statistics (Bivariate Data) — 21.
    Regression Modeling, Correlation Coefficient r,
    Line of Best Fit (The Linear Regression), Other Regressions

PART 4. Graphing Calculator

VIII. GRAPHING CALCULATOR — **23.**
1. Tips — 23.
    Clear the Memory, Return to Home Screen
2. Graph Functions — 23.
    Zoom Menu, Change Window Dimensions, Trace
3. Table of a Function — 23.
4. Calculations — 23.
    Solve Equations, Solve the System of Equations, Maximum and Minimum

REFERENCE SHEET — **24.**

PRACTICE ON REGENTS QUESTIONS — **25.**

ANSWERS TO REGENTS QUESTIONS — **60.**

Algebra 2 and Trigonometry Review

PART 1. Algebra

I. ALGEBRAIC EXPRESSIONS, EQUATIONS, AND INEQUALITIES

1. Factoring Polynomials

Find Common Factors:
$$3x^2 + 6x = 3x(x + 2)$$
$$2y^3 - 4y^2 + 2y = 2y(y^2 - 2y + 1)$$

The Difference of Two Squares:
$$a^2 - b^2 = (a + b)(a - b)$$
$$4y^2 - 25 = (2y + 5)(2y - 5)$$

Trinomial:
$$x^2 + 2x - 15 = (x + 5)(x - 3)$$
Here $5 \cdot (-3) = -15$ and $5 + (-3) = 2$

Four-term Polynomial:
If the product of the first and last terms is equal to the product of the two middle terms, the polynomial can be grouped.
e.g.
$$3x^3 - 6x^2 + 2x - 4$$
$$= 3x^2(x - 2) + 2(x - 2)$$
$$= (3x^2 + 2)(x - 2)$$

Factor Completely:
$$2x^3 - 14x^2 + 20x$$
$$= 2x(x^2 - 7x + 10) = 2x(x - 2)(x - 5)$$

2. Quadratic Equations

(1). Use Factoring

e.g. $x^2 - 10 = 3x$
$x^2 - 3x - 10 = 0$ set one side equal to zero
$(x + 2)(x - 5) = 0$ factor the trinomial
$x + 2 = 0$ or $x - 5 = 0$
$x = -2$ or $x = 5$
solution set $\{-2, 5\}$

(2). Complete the Square
$$x^2 + bx + c = 0$$
$$x^2 + bx = -c$$
$$x^2 + bx + \left(\frac{b}{2}\right)^2 = -c + \left(\frac{b}{2}\right)^2$$
$$\left(x + \frac{b}{2}\right)^2 = -c + \left(\frac{b}{2}\right)^2$$

e.g. $x^2 - 8x + 3 = 0$
$x^2 - 8x + (-4)^2 = -3 + (-4)^2$
$(x - 4)^2 = 13$
$x - 4 = \pm\sqrt{13}$
$x = 4 \pm \sqrt{13}$

(3). Quadratic Formula

$ax^2 + bx + c = 0$ where $a \neq 0$
$$x = \frac{-b \pm \sqrt{b^2 - 4ac}}{2a}$$

e.g. $2x^2 - 4x + 1 = 0$
$a = 2$, $b = -4$, $c = 1$
$$x = \frac{4 \pm \sqrt{(-4)^2 - 4(2)(1)}}{2(2)}$$
$$= \frac{4 \pm \sqrt{8}}{4} = \frac{4 \pm 2\sqrt{2}}{4}$$
$$= \frac{2 \pm \sqrt{2}}{2}$$

Quadratic Inequalities

The solution set of a quadratic inequality is in the form of

(1) $x_1 < x < x_2$ where x_1 and x_2 are the roots
or (2) $x < x_1$ or $x > x_2$ $x_1 < x_2$

Use the quadratic inequality to test a value from each of the three regions.

e.g. $x^2 - 7x + 10 < 0$
Solve the corresponding quadratic equation
$x^2 - 7x + 10 = 0$
$x_1 = 2$, $x_2 = 5$

test $x = 0$ (false), $x = 3$ (true), $x = 6$ (false);
therefore the solution set is $2 < x < 5$

e.g. $x^2 - 7x + 10 > 0$
$x_1 = 2$, $x_2 = 5$

test $x = 0$ (true), $x = 3$ (false), $x = 6$ (true);
therefore the solution set is $x < 2$ or $x > 5$

Algebra 2 and Trigonometry Review

Quadratic-Linear System

(1). Algebraic Solution:
e.g.
$$y = x^2 - 8 \quad (1)$$
$$y + 5 = 2x \quad (2)$$
From Eq. (2) $y = 2x - 5 \quad (3)$
Substitute y by $2x - 5$ in Eq.(1):
$$2x - 5 = x^2 - 8$$
$$x^2 - 2x - 3 = 0$$
$$(x - 3)(x + 1) = 0$$

$x - 3 = 0$	$x + 1 = 0$
$x = 3$	$x = -1$
$y = 2(3) - 5 = 1$	$y = 2(-1) - 5 = -7$

Solution: $\{(3, 1), (-1, -7)\}$

(2). Graphic Solution:
e.g.
$$y = x^2 - 8$$
$$y + 5 = 2x$$
rewrite in the slope-intercept form
$$y = 2x - 5$$

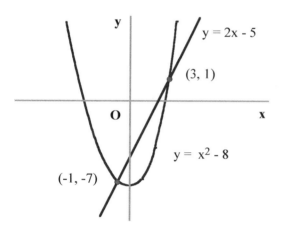

The solution of the system of equations are the intersection points (3, 1) and (-1, -7)

*Refer to **PART 4. Graphing Calculator**
Use the graphing calculator to list the table of the points and show the graphs.

3. Rational Expressions and Operations

Denominator cannot be zero.

e.g. $\dfrac{x}{x-2}$ when $x = 2$, $x - 2 = 0$, undefined

(1). Simplify: $\dfrac{2x^3}{x^2 - x - 12} \cdot \dfrac{x^2 - 16}{6x}$

$= \dfrac{2x^3(x+4)(x-4)}{(x+3)(x-4) \cdot 6x}$ factor the numerator and the denominator first;

$= \dfrac{x^2(x+4)}{3(x+3)}$ cancel out common factors in the numerator and the denominator

(2). Divide: $\dfrac{\frac{2x}{x+1}}{1 - \frac{2x}{x+1}} = \dfrac{\frac{2x}{x+1}}{\frac{x+1-2x}{x+1}} = \dfrac{\frac{2x}{x+1}}{\frac{1-x}{x+1}}$

$= \dfrac{2x}{x+1} \cdot \dfrac{x+1}{1-x} = \dfrac{2x}{1-x}$

(3). Combine: $\dfrac{1}{x+1} + \dfrac{x-1}{2}$ LCD $= 2(x+1)$

$= \dfrac{2 \cdot 1}{2(x+1)} + \dfrac{(x-1) \cdot (x+1)}{2(x+1)}$

$= \dfrac{2 + x^2 - 1}{2(x+1)} = \dfrac{x^2 + 1}{2(x+1)}$

Rational Equations

(1). Use cross-multiplication for proportions:
e.g. $\dfrac{x+2}{x-3} = \dfrac{x}{4}$
$$4(x+2) = x(x-3) \quad \text{cross-multiply}$$
$$4x + 8 = x^2 - 3x$$
$$x^2 - 7x - 8 = 0$$
$$(x-8)(x+1) = 0$$
$x = 8$ or $x = -1$ must check: Both are solutions.

(2). Use LCD:
e.g. $\dfrac{2}{x^2 - x} = \dfrac{2}{x-1} + 1$ LCD $= x(x-1)$
$$2 = 2x + x(x-1) \quad \text{multiply LCD on both sides}$$
$$2 = 2x + x^2 - x$$
$$x^2 + x - 2 = 0$$
$$(x+2)(x-1) = 0$$
$x = -2$ or $x = 1$
check: $x = -2$ ($x = 1$ undefined)

Algebra 2 and Trigonometry Review

Rational Inequalities

e.g. $\dfrac{x+2}{x-3} > \dfrac{x}{4}$

Step 1. Solve the corresponding equation
$$\dfrac{x+2}{x-3} = \dfrac{x}{4}$$
$$x = 8 \text{ or } x = -1$$

Step 2. Find the undefined values of x
$$x - 3 = 0$$
$$x = 3$$

Step 3. Divide the number line by these values

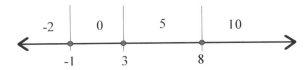

test $x = -2$ (true), $x = 0$ (false), $x = 5$ (true)
$x = 10$ (false)
thus the solution set is $x < -1$ or $3 < x < 8$
or $(-\infty, -1) \cup (3, 8)$

Note: The true and false intervals always alternate.

4. Radicals

$\sqrt{a \cdot b} = \sqrt{a} \cdot \sqrt{b}$ $a \geq 0, b \geq 0$

$\sqrt{\dfrac{a}{b}} = \dfrac{\sqrt{a}}{\sqrt{b}}$ $a \geq 0, b > 0$

Simplify: $\sqrt{75} = \sqrt{25 \cdot 3} = 5\sqrt{3}$
$\sqrt{300} = \sqrt{100 \cdot 3} = 10\sqrt{3}$

Combine: $5\sqrt{x} + 3\sqrt{x} = 8\sqrt{x}$
$\sqrt{18} - 4\sqrt{2} = \sqrt{9 \cdot 2} - 4\sqrt{2} = 3\sqrt{2} - 4\sqrt{2} = -\sqrt{2}$

Multiply: $2\sqrt{3} \cdot 4\sqrt{5} = 2 \cdot 4\sqrt{3 \cdot 5} = 8\sqrt{15}$
$3\sqrt{2} \cdot 7\sqrt{2} = 3 \cdot 7\sqrt{2 \cdot 2} = 21 \cdot 2 = 42$

Divide: $\dfrac{4\sqrt{6}}{2\sqrt{3}} = \dfrac{4}{2}\sqrt{\dfrac{6}{3}} = 2\sqrt{2}$

Rationalize: $\dfrac{1}{\sqrt{3}} = \dfrac{1}{\sqrt{3}} \cdot \dfrac{\sqrt{3}}{\sqrt{3}} = \dfrac{\sqrt{3}}{3}$

Exponents

$a^0 = 1$ $(a \neq 0)$ $5^0 = 1, (-5)^0 = 1, -5^0 = -1$

$x^{-n} = \dfrac{1}{x^n}$ $(x \neq 0)$ $5^{-2} = \dfrac{1}{5^2} = \dfrac{1}{25}$

$x^a \cdot x^b = x^{a+b}$ $5^2 \cdot 5^3 = 5^{2+3} = 5^5$

$\dfrac{x^a}{x^b} = x^{a-b}$ $\dfrac{8xy^3}{2xy} = \dfrac{8}{2} \cdot \dfrac{x}{x} \cdot \dfrac{y^3}{y} = 4y^2$

$(x^a)^b = x^{a \cdot b}$ $(5^2)^3 = 5^{2 \cdot 3} = 5^6$

Rational Exponents

$x^{\frac{a}{b}} = \sqrt[b]{x^a}$ $x^{\frac{a}{b}} = (\sqrt[b]{x})^a$ $x \geq 0$

e.g. $x^{\frac{1}{2}} = \sqrt{x}$, $x^{\frac{2}{3}} = \sqrt[3]{x^2}$

e.g. $125^{\frac{2}{3}} = (\sqrt[3]{125})^2 = 5^2 = 25$

e.g. $\dfrac{3^{\frac{1}{3}}}{3^{-\frac{2}{3}}} = 3^{\frac{1}{3} - (-\frac{2}{3})} = 3^1 = 3$

Rationalize the Denominator

e.g. $\dfrac{2}{4 + \sqrt{11}}$

$= \dfrac{2}{4 + \sqrt{11}} \cdot \dfrac{4 - \sqrt{11}}{4 - \sqrt{11}}$ multiply its conjugate

$= \dfrac{2(4 - \sqrt{11})}{16 - 11} = \dfrac{8 - 2\sqrt{11}}{5}$

e.g. $\dfrac{1}{\sqrt{3}} + \dfrac{1}{\sqrt{2}}$

$= \dfrac{1}{\sqrt{3}} \cdot \dfrac{\sqrt{3}}{\sqrt{3}} + \dfrac{1}{\sqrt{2}} \cdot \dfrac{\sqrt{2}}{\sqrt{2}}$ rationalize first

$= \dfrac{\sqrt{3}}{3} + \dfrac{\sqrt{2}}{2}$ LCD = 6

$= \dfrac{2\sqrt{3} + 3\sqrt{2}}{6}$

Algebra 2 and Trigonometry Review

Radical Equations

e.g. $x = 1 + \sqrt{x + 5}$
$x - 1 = \sqrt{x + 5}$ isolate the radical
$(x - 1)^2 = x + 5$ square both sides
$x^2 - 2x + 1 = x + 5$
$x^2 - 3x - 4 = 0$
$x = 4$ or $x = -1$
must check: $x = 4$ ($x = -1$ rejected)

e.g. $x^{\frac{3}{2}} + 1 = 9$
$x^{\frac{3}{2}} = 8$ isolate the radical
$(x^{\frac{3}{2}})^{\frac{2}{3}} = 8^{\frac{2}{3}}$
$x = 4$
must check: $x = 4$

5. Absolute Value Equations

e.g. $|4 - x| = 3x$
remove the absolute value symbol

$4 - x = 3x$	$-(4 - x) = 3x$
$x = 1$	$x = -2$
check: $|4 - 1| = 3 \cdot 1$	check: $|4 - (-2)| \neq 3 \cdot (-2)$
(true)	(false)

e.g. $|2x + 5| - 4 = x$

$(2x + 5) - 4 = x$	$-(2x + 5) - 4 = x$
$x = -1$	$x = -3$
(true)	(true)

Absolute Value Inequalities

(1) If $|x| < k$ where $k > 0$
then $-k < x < k$

e.g. $|2x + 3| < 7$
$-7 < 2x + 3 < 7$

$-7 < 2x + 3$	$2x + 3 < 7$
$-10 < 2x$	$2x < 4$
$-5 < x$	$x < 2$

solution: $-5 < x < 2$

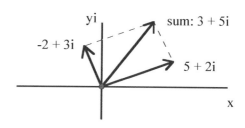

(2) If $|x| > k$ where $k > 0$
then $x < -k$ or $x > k$

e.g. $|10 - 2x| - 2 \geq 0$
$|10 - 2x| \geq 2$
$10 - 2x \leq -2$ or $10 - 2x \geq 2$
$-2x \leq -12$ or $-2x \geq -8$ divided by a negative number
$x \geq 6$ or $x \leq 4$ inequality sign reversed

6. Complex Numbers

$a + bi$ where a, b real numbers, i imaginary unit.
$a + bi = c + di$ if and only if $a = c$ and $b = d$.

e.g.

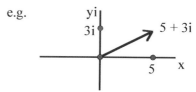

The magnitude (length) of the vector of a complex number:
e.g $|5 + 3i| = \sqrt{5^2 + 3^2} = \sqrt{34}$

Def. of the Imaginary Number:
$i = \sqrt{-1}$, $i^2 = -1$
$i^0 = 1$, $i^1 = i$, $i^2 = -1$, $i^3 = -i$
$i^{4n} = 1$, $i^{4n+1} = i$, $i^{4n+2} = -1$, $i^{4n+3} = -i$

e.g. $i^{82} = i^{4 \cdot 20 + 2} = i^2 = -1$
e.g. $\sqrt{-16} + \sqrt{-9} = 4i + 3i = 7i$
e.g. $\sqrt{-16} \cdot \sqrt{-9} = 4i \cdot 3i = 12 i^2 = -12$
(Note: $\sqrt{-16} \cdot \sqrt{-9} \neq \sqrt{(-16)(-9)} = \sqrt{144} = 12$)

Addition: $(5 + 2i) + (-2 + 3i)$

represented algebraically:
$= (5 - 2) + (2i + 3i) = 3 + 5i$

represented graphically:

Algebra 2 and Trigonometry Review

Multiplication: $(3 + i)(2 - 2i)$
$= 6 - 6i + 2i - 2i^2 = 6 - 4i + 2 = 8 - 4i$

Conjugates:
$a + bi$ and $a - bi$ are conjugates.
The sum or the product of two conjugates is a real number.
$(a + bi) + (a - bi) = 2a$
$(a + bi)(a - bi) = a^2 - (bi)^2 = a^2 - b^2 i^2 = a^2 + b^2$

Division: $\dfrac{8 + i}{2 - i}$

$= \dfrac{(8 + i)(2 + i)}{(2 - i)(2 + i)}$ $2 + i$ is the conjugate of $2 - i$

$= \dfrac{16 + 8i + 2i + i^2}{4 - i^2} = \dfrac{15 + 10i}{5} = 3 + 2i$

Multiplicative Inverse:
e.g. Write the multiplicative inverse of $2 + 4i$ in $a + bi$ form.
$\dfrac{1}{2 + 4i} = \dfrac{1}{2 + 4i} \cdot \dfrac{2 - 4i}{2 - 4i} = \dfrac{2 - 4i}{4 + 16} = \dfrac{2 - 4i}{20} = \dfrac{1}{10} - \dfrac{1}{5}i$

e.g. The multiplicative inverse of $5i$ is $\dfrac{1}{5i} = \dfrac{1}{5i} \cdot \dfrac{i}{i} = \dfrac{i}{-5} = -\dfrac{i}{5}$

7. Roots of the Quadratic Equations

$ax^2 + bx + c = 0$ where $a \neq 0$

The sum of the roots $x_1 + x_2 = -\dfrac{b}{a}$

The product of the roots $x_1 \cdot x_2 = \dfrac{c}{a}$

e.g. $x_1 + x_2 = 5$ and $x_1 \cdot x_2 = 6$
 Write the quadratic equation.
 $5 = -\dfrac{b}{a}$, $6 = \dfrac{c}{a}$
 $a = 1$, $b = -5$, $c = 6$
 $x^2 - 5x + 6 = 0$

$b^2 - 4ac$ is called discriminant Δ
$\Delta > 0$ two unequal real roots (two x intercepts)
$\Delta = 0$ two equal real roots (one x intercept)
$\Delta < 0$ no real roots (no x intercept)
when $\Delta < 0$, it has two imaginary roots in the form of conjugates: $x_1 = a + bi$ and $x_2 = a - bi$
e.g. $x^2 + 2x + 2 = 0$ $x_1 = -1 + i$, $x_2 = -1 - i$

8. Polynomial Equations of Higher Degrees

A polynomial equation of degree n has n roots which are real, imaginary, or both.
The real roots are the x-intercepts and the imaginary roots are paired complex conjugates.

e.g $x^3 - 3x^2 + 4x - 12 = 0$
 $x^2(x - 3) + 4(x - 3) = 0$
 $(x - 3)(x^2 + 4) = 0$

$x - 3 = 0$ | $x^2 + 4 = 0$
$x = 3$ | $x^2 = -4$
 | $x = \pm \sqrt{-4}$
 | $x = \pm 2i$

Solution: $\{3, -2i, 2i\}$

e.g. $x^4 + 3x^2 = 4$
 $x^4 + 3x^2 - 4 = 0$
 $(x^2 + 4)(x^2 - 1) = 0$

$x^2 + 4 = 0$ | $x^2 - 1 = 0$
$x = \pm 2i$ | $(x + 1)(x - 1) = 0$
 | $x = -1$, $x = 1$

Solution set: $\{-1, 1, -2i, 2i\}$

9. Sequence and Series

(1). A **sequence** is a special type of function:
 $a_n = f(n)$ where $\{n: 1, 2, 3, 4, 5, \cdots\}$

Recursive Definition:
 In some sequences, except the first one, any term can be defined by its previous terms.

e.g. sequence $\{2, 4, 8, 16, \cdots\}$
 $a_n = 2^n$
 use recursive definition:
 $a_n = 2\, a_{n-1}$ for $n > 1$

A **series** is the sum of the sequence.

e.g. $S_5 = 2 + 4 + 8 + 16 + 32$

Algebra 2 and Trigonometry Review

(2). Sigma Notation Σ

e.g. $\sum_{i=1}^{5} i = 1 + 2 + 3 + 4 + 5$

here i is the index, lower limit = 1, upper limit = 5

e.g. $\sum_{n=1}^{5} (2n - 1) = 1 + 3 + 5 + 7 + 9$

e.g. $\sum_{k=3}^{7} k^2 = 3^2 + 4^2 + 5^2 + 6^2 + 7^2$

(3). Arithmetic Sequences

$a_n = a_1 + (n - 1)d$ d is the common difference

or $a_n = a_{n-1} + d$ recursive definition

e.g. { 5, 10, 15, 20, 25 }

e.g. Find the missing terms between $a_1 = 5$ and $a_6 = 15$ in an arithmetic sequence.
$a_n = a_1 + (n - 1)d$
$a_6 = a_1 + (6 - 1)d$
$15 = 5 + (6 - 1)d$
$d = 2$
$a_2 = 7, a_3 = 9, a_4 = 11, a_5 = 13$

These numbers are called **arithmetic means** between a_1 and a_6.

Arithmetic Series

The sum of the first n terms:

$$S_n = \sum_{i=1}^{n} a_i = \frac{n}{2}(a_1 + a_n)$$

e.g. $1 + 2 + 3 + 4 + \cdots + 100$
$= \frac{100}{2}(1 + 100) = 5050$

(4). Geometric Sequences

$a_n = a_1 \cdot r^{n-1}$ r is the common ratio

or $a_n = a_{n-1} \cdot r$ recursive definition

e.g. $\{ 1, \frac{1}{2}, \frac{1}{4}, \frac{1}{8}, \frac{1}{16} \}$

e.g. $\{ 2, -4, 8, -16, 32, \cdots \}$

e.g. Find the **geometric means** between $a_1 = 5$ and $a_5 = 80$
$a_n = a_1 \cdot r^{n-1}$
$a_5 = a_1 \cdot r^{5-1}$
$80 = 5 \cdot r^4$
$r^4 = 16$
$r = 2$ or $r = -2$
$a_2 = 10, a_3 = 20, a_4 = 40$
or $a_2 = -10, a_3 = 20, a_4 = -40$

Geometric Series

Since $\frac{(1 - r^n)}{1 - r} = 1 + r + r^2 + \cdots + r^{n-1}$

The sum of the first n terms:

$$S_n = \sum_{i=1}^{n} a_i = a_1 + a_1 r + a_1 r^2 + \cdots + a_1 r^{n-1} = \frac{a_1(1 - r^n)}{1 - r}$$

e.g. $5 + \frac{5}{2} + \frac{5}{4} + \frac{5}{8} + \frac{5}{16}$

$a_1 = 5, r = \frac{1}{2}, n = 5$

$$S_n = \frac{a_1(1 - r^n)}{1 - r} = \frac{5(1 - (\frac{1}{2})^5)}{(1 - \frac{1}{2})} = \frac{5 \cdot \frac{31}{32}}{\frac{1}{2}} = \frac{155}{16}$$

(5). Infinite Series

1. An infinite arithmetic series has no limit.
2. An infinite geometric series has no limit when $|r| \geq 1$.
3. An infinite geometric series has a limit when $|r| < 1$.

$$S_n = \sum_{i=1}^{\infty} a_i = \frac{a_1}{1 - r}$$

e.g. Find the limit of $0.252525 \cdots$
$0.252525 \cdots = 0.25 + 0.0025 + 0.000025 + \cdots$
$a_1 = 0.25, r = \frac{1}{100} = 0.01$

$$S_n = \frac{a_1}{1 - r} = \frac{0.25}{1 - 0.01} = \frac{0.25}{0.99} = \frac{25}{99}$$

4. The Number e

$$\sum_{n=0}^{\infty} \frac{1}{n!} = 1 + \frac{1}{1!} + \frac{1}{2!} + \frac{1}{3!} + \cdots = e$$

$e = 2.718281828 \cdots$

II. RELATIONS AND FUNCTIONS

1. Functions
Each element x in the domain is assigned to exactly one element y in the range.

Relation (Not a Function)

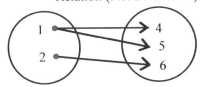

Function

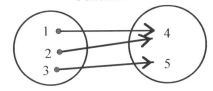

One-to-One Function

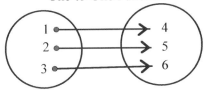

Domain and Range

e.g. $y = x^2$ Domain: $\{x \mid x \text{ all real numbers}\}$
                 Range: $\{y \mid y \geq 0\}$

e.g. $y = \sqrt{x}$ Domain: $\{x \mid x \geq 0\}$
                 Range: $\{y \mid y \geq 0\}$

e.g. $y = \dfrac{1}{x^2 - 9}$ Domain: $\{x \mid x \text{ all reals except} \pm 3\}$

e.g. $y = \dfrac{1}{\sqrt{x - 3}}$ Domain: $\{x \mid x > 3\}$

Interval Notation

e.g. $(2, 5)$ represents $\{x \mid 2 < x < 5\}$
      $[2, 5]$ represents $\{x \mid 2 \leq x \leq 5\}$
      $(2, 5]$ represents $\{x \mid 2 < x \leq 5\}$
      $[2, 5)$ represents $\{x \mid 2 \leq x < 5\}$

e.g. $(-\infty, \infty)$ represents $\{x \mid x \text{ all real numbers}\}$
      $(-\infty, -5)$ represents $\{x \mid x < -5\}$
      $[5, \infty)$ represents $\{x \mid x \geq 5\}$
      $(-\infty, -5) \cup [5, \infty)$ represents $\{x < -5 \text{ or } x \geq 5\}$

Function Notation

e.g. $y = \dfrac{2x}{x - 1}$, its function notation: $f(x) = \dfrac{2x}{x - 1}$

then $f(5) = \dfrac{2 \cdot 5}{5 - 1} = \dfrac{10}{4} = \dfrac{5}{2}$

$f(a + 2) = \dfrac{2(a + 2)}{(a + 2) - 1} = \dfrac{2a + 4}{a + 1}$

$f(x^2) = \dfrac{2(x^2)}{(x^2) - 1} = \dfrac{2x^2}{x^2 - 1}$

Vertical Line Test:
If any vertical line intersects the graph at only one point, then the relation is a function.

e.g. $y^2 = x$ is equivalent to $y = \pm\sqrt{x}$. It is not a function.

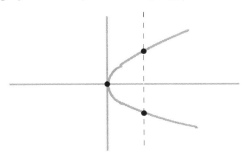

Horizontal Line Test:
If any horizontal line intersects the graph at only one point, then the function is a **one-to-one function**.

e.g. $y = x^2$ is a function, but not a one-to-one function.

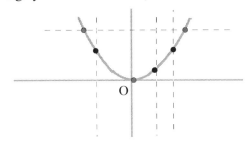

If we restrict the domain of $y = x^2$ to $x \geq 0$, then the function is a one-to-one funcion.

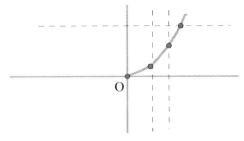

Onto

A function from set A to set B is onto if the range of the function is equal to set B.

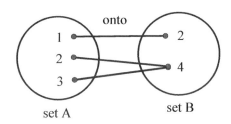

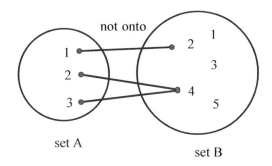

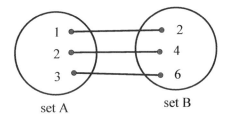

e.g. Set A and set B both are all real numbers.
 $y = x^3$ is onto and one-to-one.
 $y = x^2$ is neither onto nor one-to-one.
(the range $y \geq 0$ is not equal to all real numbers)
(not passing the horizontal line test).

e.g. Set A and set B both are all real numbers.
 $y = \sqrt{x}$ is one-to-one function but not onto.

2. Composition of Functions

$$(f \circ g)(x) = f(g(x))$$

e.g. $f(x) = x^2 - 1$, $g(x) = x + 1$
 $(f \circ g)(x) = f(g(x)) = f(x+1) = (x+1)^2 - 1 = x^2 + 2x$
 $(f \circ g)(2) = f(g(2)) = f(2+1) = f(3) = 3^2 - 1 = 8$
 $(g \circ f)(x) = g(f(x)) = g(x^2 - 1) = (x^2 - 1) + 1 = x^2$
 $(g \circ f)(2) = g(f(2)) = g(2^2 - 1) = g(3) = 3 + 1 = 4$
Note: $(f \circ g)(x) \neq (g \circ f)(x)$

3. Inverse Functions

For every one-to-one function f(x), there is an inverse function $f^{-1}(x)$. (passing both vertical and horizontal line tests)

e.g. $y = x^2$
It is a function. (passing the vertical line test)
But it is not a one-to-one function. (fails the horizonal line test)
The domain of the inverse function is the range of the original function.

e.g. Original f(x) = {(1, 1), (2, 4), (3, 9)}
 Domain: {x | x = 1, 2, 3}
 Range: {y | y = 1, 4, 9}
 Inverse $f^{-1}(x)$ = {(1, 1), (4, 2), (9, 3)}
 Domain: {x | x = 1, 4, 9}
 Range: {y | y = 1, 2, 3}

e.g. f(x) = 3x + 5   find $f^{-1}(x)$
 y = 3x + 5   express $f^{-1}(x)$ as y
 x = 3y + 5   interchange x and y
 $y = \dfrac{x - 5}{3}$ solve for y in terms of x
 $f^{-1}(x) = \dfrac{x - 5}{3}$

The graph of $f^{-1}(x)$ is the reflection of f(x) in the y = x.

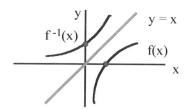

Algebra 2 and Trigonometry Review

4. Functions under a Transformation

Translation:
$y = f(x)$ ___T_{a,b}___ $y = f(x - a) + b$

Reflection:
$y = f(x)$ ___r_{x-axis}___ $y = -f(x)$

$y = f(x)$ ___r_{y-axis}___ $y = f(-x)$

Dilation:
$y = f(x)$ ___vertical stretch if $a > 1$___ $y = af(x)$
        vertical shrink if $0 < a < 1$

The transformation rules for functions are different from the transformation rules for images.
Tips: Use the graphing calculator to verify the answer.

e.g.

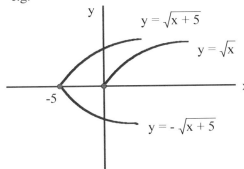

$y = \sqrt{x}$ ___moved 5 units to the left___ $y = \sqrt{x + 5}$

$y = \sqrt{x + 5}$ ___reflected in the x-axis___ $y = -\sqrt{x + 5}$

e.g.

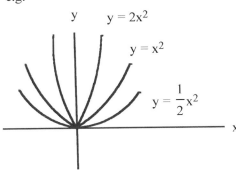

$y = x^2$ ___stretched vertically by a factor of 2___ $y = 2x^2$

$y = x^2$ ___shrunk vertically by a factor of 1/2___ $y = \frac{1}{2}x^2$

5. Important Functions and Relations

(1). Direct Variation

A straight line passing through the Origin

$$y = mx \quad \text{or} \quad \frac{y}{x} = m$$

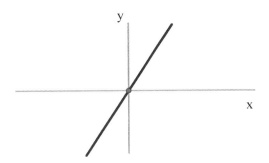

To solve a problem, use $\frac{x_1}{x_2} = \frac{y_1}{y_2}$

e.g. The distance varies directly as the time when a car travels at a constant speed.
$$d = s \cdot t \quad (\text{s is a constant})$$

(2). Inverse Variations: (Function)

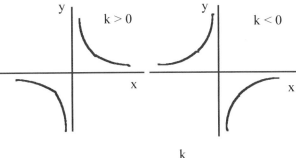

$$xy = k \quad \text{or} \quad y = \frac{k}{x}$$

To solve a problem, use $x_1 \cdot y_1 = x_2 \cdot y_2$

e.g. The speed varies inversely to the time when a car travels over a certain distance.
$$s \cdot t = d \quad (\text{d is a constant})$$

Algebra 2 and Trigonometry Review

(3). Absolute Value Functions
$$y = |x|$$

when $x < 0$, $y = -x$

when $x \geq 0$, $y = x$

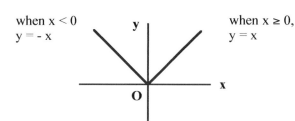

e.g. $y = |2x - 4|$

when $2x - 4 < 0$
$x < 2$
$y = -(2x - 4)$
$y = -2x + 4$

when $2x - 4 \geq 0$
$x \geq 2$
$y = 2x - 4$

(4). Quadratic Functions and Parabolas

General Form:
$$y = f(x) = ax^2 + bx + c \quad \text{where } a \neq 0$$

(1). Axis of Symmetry: $x = -\dfrac{b}{2a}$

(2). Vertex (Turning Point): (x, y)
$$x = -\dfrac{b}{2a}, \quad y = f(x) = f\left(-\dfrac{b}{2a}\right)$$

(3). Opening:

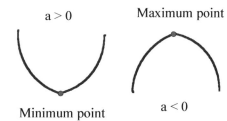

$a > 0$ Minimum point

$a < 0$ Maximum point

e.g. $y = 12x - 2x^2$
$y = -2x^2 + 12x$ write in standard form
$a = -2, \ b = 12, \ c = 0$

(1). Axis of Symmetry: $x = -\dfrac{b}{2a} = -\dfrac{12}{2(-2)} = 3$

(2). Vertex: $x = 3, \ y = -2(3)^2 + 12(3) = 18$
    (3, 18)

(3). Opening: $a = -2 < 0$
  It has a maximum of 18 at $x = 3$.

Solve Quadratic Inequalities in Two Variables

e.g. $x^2 - 2x < 5 + y$
Rewrite it in the standard form:
$$y > x^2 - 2x - 5$$
Graph $y = x^2 - 2x - 5$ in dashed line.
The shaded region above the curve is the solution

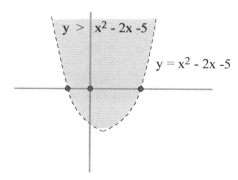

(5). Equations of Circles (Relation, not Function)

The center-radius equation of a circle with radius r and center (h, k)

$$(x - h)^2 + (y - k)^2 = r^2$$

e.g. $x^2 + y^2 + 4x - 6y - 12 = 0$
Find its center and radius, and graph it.
$$x^2 + 4x + y^2 - 6y = 12$$
Complete the squares
$$x^2 + 4x + \left(\dfrac{4}{2}\right)^2 + y^2 - 6y + \left(\dfrac{-6}{2}\right)^2$$
$$= 12 + \left(\dfrac{4}{2}\right)^2 + \left(\dfrac{-6}{2}\right)^2$$
$$x^2 + 4x + 4 + y^2 - 6y + 9 = 12 + 4 + 9$$
$$(x + 2)^2 + (y - 3)^2 = 5^2$$
center is $(-2, 3)$ and the radius is 5

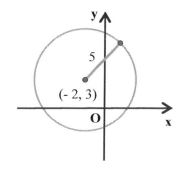

Algebra 2 and Trigonometry Review

(6). Exponential Functions and Equations

$$y = a^x \quad \text{where } a > 0 \text{ and } a \neq 1$$

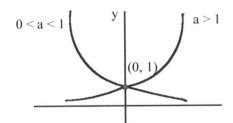

(1). Domain: $\{ x \mid x \text{ all real numbers} \}$
    Range: $\{ y \mid y > 0 \}$
(2). $a > 1$, the function is increasing (exponential growth);
    $a < 1$, the function is decreasing (exponential decay).
(3). when $x = 0$, $y = 1$ (the graphs of the exponential functions pass point $(0, 1)$).
(4). x-axis is the horizontal asymptote.
(5). $y = (\frac{1}{a})^x$ and $y = a^x$ are symmetric in the y-axis.

General Form:

$$y = k \cdot a^x \quad \text{where } a > 0 \text{ and } a \neq 1, k \text{ is a constant}$$

Exponential Models:

e.g. $A = A_0 e^{-0.025t}$

A_0 is the original amount;
A is the amount at time t;
Find the half-life of this exponential decay.

half-life: $A = \frac{1}{2} A_0 = 0.5 A_0$

$0.5 A_0 = A_0 e^{-0.025t}$
$0.5 = e^{-0.025t}$
$\ln 0.5 = -0.025t \ln e \quad (\ln e = 1)$
$t = \dfrac{\ln 0.5}{-0.025} = 27.726$

Exponential Equations

e.g. $\quad 9^{x+1} = 27^x$
$\quad\quad 3^{2(x+1)} = 3^{3x} \quad$ transform to same base
$\quad\quad 2(x+1) = 3x$
$\quad\quad\quad x = 2 \quad$ check: true

e.g. $\quad 3^x = 5$
$\quad\quad \log 3^x = \log 5$
$\quad\quad x \log 3 = \log 5$
$\quad\quad\quad x = \dfrac{\log 5}{\log 3} = 1.465$

(7). Logarithmic Functions and Equations

$y = \log_a x$ is equivalent to $x = a^y$
e.g. $\log_5 25 = 2$ is equivalent to $5^2 = 25$

$y = \log_a x$ is the inverse function of $y = a^x$

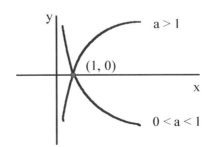

Domain: $\{ x \mid x > 0 \}$
Range: $\{ y \mid y \text{ all real numbers} \}$

$\log_a 1 = 0, \quad \log_a a = 1$
$\log_a AB = \log_a A + \log_a B$
$\log_a \dfrac{A}{B} = \log_a A - \log_a B$
$\log_a A^n = n \cdot \log_a A, \quad \log_a \sqrt[n]{A} = \dfrac{1}{n} \log_a A$

e.g. $\log_2 \sqrt[3]{2} = \log_2 2^{\frac{1}{3}} = \dfrac{1}{3} \log_2 2 = \dfrac{1}{3}$

e.g. If $\log_5 x = 2$, what is the value of $\sqrt{x}$?
$\log_5 x = 2$ is equivalent to $x = 5^2 = 25$
$\sqrt{x} = \sqrt{25} = 5$

Common Logarithms

$$\log A = \log_{10} A$$

e.g. $\log 100 = \log 10^2 = 2$
e.g. $\log 0.01 = \log 10^{-2} = -2$
e.g. $\log 123 = \log 1.23 \cdot 10^2 = 2 + \log 1.23$
e.g. If $\log A = 2$ and $\log B = 3$, then
$\log \dfrac{\sqrt{A}}{B^3} = \dfrac{1}{2} \log A - 3 \log B = \dfrac{1}{2} \cdot 2 - 3 \cdot 3 = -8$

e.g. $\log_2 5 = \dfrac{\log 5}{\log 2} \approx 2.32$

Natural Logarithms

$$\ln A = \log_e A$$

$$e = \sum_{n=0}^{\infty} \frac{1}{n!} = 1 + \frac{1}{1!} + \frac{1}{2!} + \frac{1}{3!} + \cdots$$

$$e = 2.718281828 \cdots$$

$\ln 10 \approx 2.3 \qquad \log e \approx 0.4343$

e.g. $\ln x = 5$
$x = e^5 \approx 148.4$

e.g. $e^{2\ln 5} = e^{\ln 5^2} = 5^2 = 25$

Change of Base Formula

$$\log_a A = \frac{\log A}{\log a}$$

or $\log_a A = \dfrac{\ln A}{\ln a}$

e.g. $\log_2 5 = \dfrac{\log 5}{\log 2} = 2.3219$

Logarithmic Equations

e.g. $\log_x 4 + \log_x 9 = 2$
$\log_x 4 \cdot 9 = 2$
$x^2 = 36$
$x = 6 \qquad (x = -6 \text{ rejected})$

e.g. $\log_4(x^2 + 3x) - \log_4(x + 5) = 1$
$\log_4 \dfrac{x^2 + 3x}{x + 5} = \log_4 4$
$\dfrac{x^2 + 3x}{x + 5} = 4$
$x^2 + 3x = 4x + 20$
$x^2 - x - 20 = 0$
$(x + 4)(x - 5) = 0$
$x = -4 \text{ or } x = 5 \qquad \text{check: true}$

PART 2. Trigonometry

III. TRIGONOMETRIC FUNCTIONS

1. Degrees and Radians

The degree measure of a circle is 360.
The degree measure of a semicircle is 180.
$\qquad 1° = 60'$ (minutes)
e.g. $\quad 0.3° = 0.3 \cdot 60 = 18'$
$\qquad 45' = \dfrac{45}{60} = 0.75°$

Radian is a different unit to measure the angle and the arc.
The radian measure of a circle is 2π.
The radian measure of a semicircle is π.
$\quad 2\pi \text{ (radians)} = 360°$
$\quad \pi \text{ (radians)} = 180°$
$\quad 1 \text{ (radian)} = \dfrac{180°}{\pi} \approx 57.3°$

Conversion:

use $\quad \dfrac{\pi}{180°} = 1 \quad$ or $\quad \dfrac{180°}{\pi} = 1$

e.g. $\quad 60° = 60° \cdot \dfrac{\pi}{180°} = \dfrac{\pi}{3}$

e.g. $\quad \dfrac{\pi}{4} = \dfrac{\pi}{4} \cdot \dfrac{180°}{\pi} = 45°$

Coterminal Angles:

Coterminal angles have the same terminal side.
They differ 360° or a multiple of 360°.

e.g. 30°, 390°, and 750° are coterminal angles.

e.g. -60° and 300° are coterminal angles.

Arc Length:
$\quad s = r \cdot \theta \qquad$ where r is radius and θ in radian

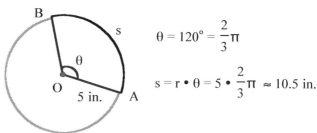

$\theta = 120° = \dfrac{2}{3}\pi$

$s = r \cdot \theta = 5 \cdot \dfrac{2}{3}\pi \approx 10.5 \text{ in.}$

These problems can also be solved by using fractions or ratios to the circle.

Algebra 2 and Trigonometry Review

2. Trigonometric Fuctions

Trigonometric Ratios and Basic Functions

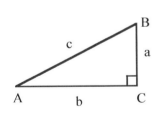

$$\sin A = \frac{Opp}{Hyp} = \frac{a}{c}$$

$$\cos A = \frac{Adj}{Hyp} = \frac{b}{c}$$

$$\tan A = \frac{Opp}{Adj} = \frac{a}{b}$$

Reciprocal Functions:

$$\cot A = \frac{1}{\tan A} \; , \; \sec A = \frac{1}{\cos A} \; , \; \csc A = \frac{1}{\sin A}$$

Exact Values to Remember:

θ (degree)	0°	30°	45°	60°	90°
θ	0	π/6	π/4	π/3	π/2
sinθ	0	1/2	$\sqrt{2}/2$	$\sqrt{3}/2$	1
cosθ	1	$\sqrt{3}/2$	$\sqrt{2}/2$	1/2	0
tanθ	0	$\sqrt{3}/3$	1	$\sqrt{3}$	undefined

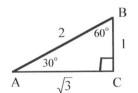 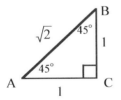

Pythagorean Triples:
3, 4, 5 and 5, 12, 13

e.g. If $\sin\theta = \frac{3}{5}$, then $\cos\theta = \frac{4}{5}$, $\tan\theta = \frac{3}{4}$

If $\tan\theta = \frac{5}{12}$, then $\sin\theta = \frac{5}{13}$, $\cos\theta = \frac{12}{13}$

Cofunctions

$\sin\theta = \cos(90° - \theta)$,
$\tan\theta = \cot(90° - \theta)$
$\sec\theta = \csc(90° - \theta)$

e.g. $\sin 30° = \cos 60°$
e.g. If $\sin A = \cos B$, then $A + B = 90°$
e.g. If $\sin 2A = \cos 4A$, find A
  $2A + 4A = 90°$ $A = 15°$

Unit Circle

Center at the origin ; Radius of one unit.

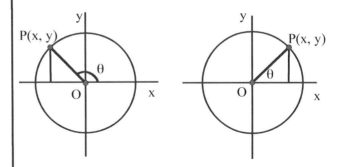

For Angles of Any Degree:

$$\cos\theta = x \; , \; \sin\theta = y \; , \; \tan\theta = \frac{y}{x}$$

In general, a point is not on a Unit Circle:

$$\cos\theta = \frac{x}{\sqrt{x^2 + y^2}} \; , \; \sin\theta = \frac{y}{\sqrt{x^2 + y^2}} \; , \; \tan\theta = \frac{y}{x}$$

Reference Angle

An acute angle formed by the terminal side of the given angle and the x-axis.

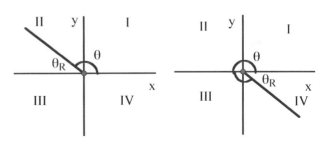

Quadrant	I	II	III	IV
θ_R	θ	180° − θ	θ − 180°	360° − θ
sinθ	$\sin\theta_R$	$\sin\theta_R$	$-\sin\theta_R$	$-\sin\theta_R$
cosθ	$\cos\theta_R$	$-\cos\theta_R$	$-\cos\theta_R$	$\cos\theta_R$
tanθ	$\tan\theta_R$	$-\tan\theta_R$	$\tan\theta_R$	$-\tan\theta_R$

e.g. $\cos 120° = -\cos 60° = -\frac{1}{2}$ ($\theta_R = 180° - 120° = 60°$)

$\cos 240° = -\cos 60° = -\frac{1}{2}$ ($\theta_R = 240° - 180° = 60°$)

$\cos 300° = \cos 60° = \frac{1}{2}$ ($\theta_R = 360° - 300° = 60°$)

IV. TRIGONOMETRIC GRAPHS

* Set the Graphing Calculator in Radian Mode

1. Graphs of Trigonometric Functions

$y = \sin x$

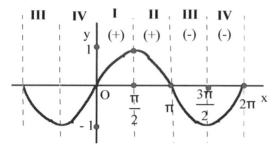

Domain: { x | x all real numbers }
Range: { y | -1 ≤ y ≤ 1 }

$y = \cos x$

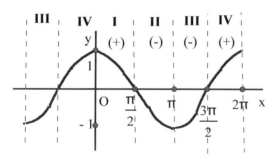

Domain: { x | x all real numbers }
Range: { y | -1 ≤ y ≤ 1 }

$y = \tan x$

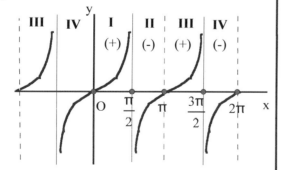

Domain: $\{ x \mid x \neq \frac{\pi}{2} + n\pi \text{ for n an integer} \}$
Range: { y | y all real numbers }

2. Graphs of the Reciprocal Functions

when $f(x) \to 0$, $\frac{1}{f(x)} \to \infty$

$0 < f(x) < 1$, $\frac{1}{f(x)} > 1$

$f(x) = 1$, $\frac{1}{f(x)} = 1$

$f(x) > 1$, $\frac{1}{f(x)} < 1$

$f(x) \to \infty$, $\frac{1}{f(x)} \to 0$, etc.

e.g. $y = \csc x = \frac{1}{\sin x}$

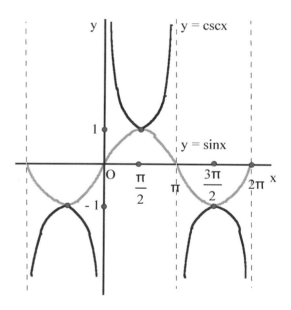

$y = \csc x$
Domain: { x | x : all real numbers except nπ }
Range: { y | y ≤ -1 or y ≥ 1 }

Use graphing calculator to see $y = \cot x$ and $y = \sec x$.

Algebra 2 and Trigonometry Review

3. Amplitude, Period, and Frequency:

$y = a\sin bx$ and $y = a\cos bx$

amplitude $= |a|$, frequency $= |b|$, period $= \dfrac{2\pi}{|b|}$

e.g. $y = \cos x$
amplitude $= 1$, frequency $= 1$, period $= 2\pi$

e.g. $y = -3\sin 2x$
amplitude $= 3$, frequency $= 2$, period $= \dfrac{2\pi}{2} = \pi$

$y = \tan bx$

frequency $= |b|$, period $= \dfrac{\pi}{|b|}$

e.g. $y = \tan x$ period $= \pi$
e.g. $y = \tan 2x$ period $= \dfrac{\pi}{2}$

The graph of $y = a\sin b(x + c) + d$

e.g. Graph $y = 5\sin 2(x + \dfrac{\pi}{4}) + 3$

Amplitude $= 5$, Frequency $= 2$, Period $= \pi$

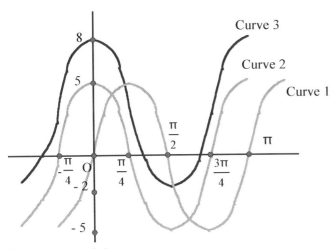

Curve 1: $y = 5\sin 2x$
Curve 2: $y = 5\sin 2(x + \dfrac{\pi}{4})$ shift $\dfrac{\pi}{4}$ to the left (phase shift $-\dfrac{\pi}{4}$)
Curve 3: $y = 5\sin 2(x + \dfrac{\pi}{4}) + 3$ shift 3 units up

4. Inverse Trigonometric Functions

$y = \arcsin x$ or $y = \sin^{-1} x$

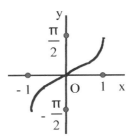

Domain: $\{ x \mid -1 \leq x \leq 1 \}$

Range: $\{ y \mid -\dfrac{\pi}{2} \leq y \leq \dfrac{\pi}{2} \}$

$y = \arccos x$ or $y = \cos^{-1} x$

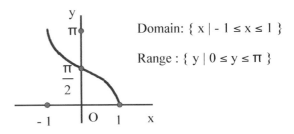

Domain: $\{ x \mid -1 \leq x \leq 1 \}$

Range: $\{ y \mid 0 \leq y \leq \pi \}$

$y = \arctan x$ or $y = \tan^{-1} x$

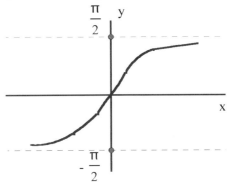

Domain: $\{ x \mid x \text{ all real numbers} \}$
Range: $\{ y \mid -\dfrac{\pi}{2} < y < \dfrac{\pi}{2} \}$

e.g. Find $\sin(\arccos \dfrac{1}{2})$

Let $\theta = \arccos \dfrac{1}{2}$ $(0 \leq \theta \leq \pi)$

$\cos \theta = \dfrac{1}{2}$

since $0 \leq \theta \leq \pi$, $\theta = 60°$

$\sin \theta = \dfrac{\sqrt{3}}{2}$

V. TRIGONOMETRIC APPLICATIONS

1. Trigonometric Identities

$$\tan\theta = \frac{\sin\theta}{\cos\theta}, \quad \cot\theta = \frac{\cos\theta}{\sin\theta}$$

$$\sin^2\theta + \cos^2\theta = 1$$
$$\tan^2\theta + 1 = \sec^2\theta$$
$$\cot^2\theta + 1 = \csc^2\theta$$

e.g. $\sin\theta = \frac{5}{13}$ and θ is in Quadrant II.

Find the value of (1) $\cos\theta$ and (2) $\tan\theta$.

(1) $(\frac{5}{13})^2 + \cos^2\theta = 1$

$\cos\theta = -\frac{12}{13}$ ($\cos\theta$ is negative in Q II)

(2) $\tan\theta = \frac{\sin\theta}{\cos\theta} = \frac{5/13}{-12/13} = -\frac{5}{12}$

Two Angles

$\sin(A + B) = \sin A\cos B + \cos A\sin B$
$\sin(A - B) = \sin A\cos B - \cos A\sin B$
$\cos(A + B) = \cos A\cos B - \sin A\sin B$
$\cos(A - B) = \cos A\cos B + \sin A\sin B$

e.g. The following identities are true for all values of θ:
$\sin(-\theta) = -\sin\theta$, $\cos(-\theta) = \cos\theta$, $\tan(-\theta) = -\tan\theta$
$\sin(90° - \theta) = \cos\theta$, $\cos(90° - \theta) = \sin\theta$

Double Angles

$\sin 2A = 2\sin A\cos A$
$\cos 2A = \cos^2 A - \sin^2 A$
$\quad\quad\;\; = 2\cos^2 A - 1$
$\quad\quad\;\; = 1 - 2\sin^2 A$
$\tan 2A = \frac{2\tan A}{1 - \tan^2 A}$

Half Angles

$\sin\frac{1}{2}A = \pm\sqrt{\frac{1 - \cos A}{2}}$

$\cos\frac{1}{2}A = \pm\sqrt{\frac{1 + \cos A}{2}}$

$\tan\frac{1}{2}A = \pm\sqrt{\frac{1 - \cos A}{1 + \cos A}}$

Find θ in different Quadrants in terms of θ_R

Quadrant	I	II	III	IV
θ	θ_R	$180° - \theta_R$	$180° + \theta_R$	$360° - \theta_R$

e.g. Find the exact value of $\cos 105°$
$\cos 105° = \cos(60° + 45°)$
$\quad\quad\;\;\; = \cos 60°\cos 45° - \sin 60°\sin 45°$
$\quad\quad\;\;\; = \frac{1}{2}\cdot\frac{\sqrt{2}}{2} - \frac{\sqrt{3}}{2}\cdot\frac{\sqrt{2}}{2}$
$\quad\quad\;\;\; = \frac{\sqrt{2} - \sqrt{6}}{4}$

e.g. $\sin x = \frac{4}{5}$, x is an acute angle.
Find the value of $\cos 2x$ and $\sin 2x$.

(1) $\cos 2x = 1 - 2\sin^2 x$
$\quad\quad\;\;\; = 1 - 2(\frac{4}{5})^2 = -\frac{7}{25}$

(2) $\cos x = \sqrt{1 - \sin^2 x} = \frac{3}{5}$ (acute angle)

$\sin 2x = 2\sin x\cos x$
$\quad\quad\;\; = 2(\frac{4}{5})(\frac{3}{5}) = \frac{24}{25}$

or $\sin 2x = \sqrt{1 - \cos^2 2x} = \frac{24}{25}$

(sine is positive when 2x is in either Quadrant I or II,)

2. Trigonometric Equations

e.g. Solve for θ, $0° \leq \theta < 180°$
$2\cos^2\theta - \sin\theta = 1$
$2(1 - \sin^2\theta) - \sin\theta = 1$
$2 - 2\sin^2\theta - \sin\theta = 1$
$2\sin^2\theta + \sin\theta - 1 = 0$
$(2\sin\theta - 1)(\sin\theta + 1) = 0$

$2\sin\theta - 1 = 0$	$\sin\theta + 1 = 0$
$\sin\theta = \frac{1}{2}$	$\sin\theta = -1$
$\theta = 30°$	$\theta = 270°$
	(not in the range)

θ in Quadrant I is also the reference angle θ_R.
Another angle in Quadrant II, $\theta = 180 - \theta_R = 150°$
{ 30°, 150° }

e.g. Solve for θ, $0° \leq \theta < 360°$
$\cos 2\theta - \sin\theta = 1$
$1 - 2\sin^2\theta - \sin\theta = 1$
$\sin\theta(2\sin\theta + 1) = 0$

$\sin\theta = 0$	$2\sin\theta + 1 = 0$
$\theta = 0°$ or $\theta = 180°$	$\sin\theta = -0.5$ no θ in Quadrant I
	Solve $\sin\theta_R = 0.5$, $\theta_R = 30°$
	$\theta = 180° + 30° = 210°$, or
	$\theta = 360° - 30° = 330°$

{ 0°, 180°, 210°, 330°}

Algebra 2 and Trigonometry Review

3. Applications

For any triangle:

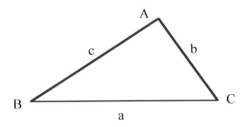

$$\text{Area} = \frac{1}{2}ab\sin C = \frac{1}{2}bc\sin A = \frac{1}{2}ca\sin B$$

Law of Sines:

$$\frac{a}{\sin A} = \frac{b}{\sin B} = \frac{c}{\sin C}$$

Law of Cosines:

$$c^2 = a^2 + b^2 - 2ab\cos C$$
$$\cos C = \frac{a^2 + b^2 - c^2}{2ab}$$

e.g. Find the area of $\triangle ABC$

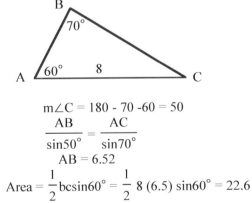

$m\angle C = 180 - 70 - 60 = 50$
$\frac{AB}{\sin 50°} = \frac{AC}{\sin 70°}$
$AB = 6.52$
$\text{Area} = \frac{1}{2}bc\sin 60° = \frac{1}{2}\,8\,(6.5)\sin 60° = 22.6$

e.g. **Ambiguous Case:**

In $\triangle ABC$, $m\angle A = 30$, $AB = 12$, $BC = 7$,
How many possible trianges can be constructed?

$$\frac{BC}{\sin A} = \frac{AB}{\sin C}$$
$$\frac{7}{\sin 30°} = \frac{12}{\sin C}$$
$$\sin C = \frac{6}{7}$$

$m\angle C = 59$ or $m\angle C = 180 - 59 = 121$
check: $m\angle A + m\angle C = 30 + 59 < 180$ OK
check: $m\angle A + m\angle C = 30 + 121 < 180$ OK
Two possible triangles can be constructed.

e.g. Find x.

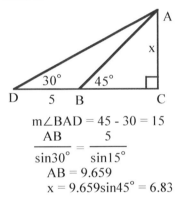

$m\angle BAD = 45 - 30 = 15$
$\frac{AB}{\sin 30°} = \frac{5}{\sin 15°}$
$AB = 9.659$
$x = 9.659\sin 45° = 6.83$

e.g. Two forces of 10 lb. and 15 lb. act on a body. Their resultant is 20 lb. Find the angle between the two applied forces.

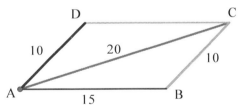

$BC = AD = 10$ (in a parallelogram opposite sides $\cong$)
In $\triangle ABC$, $\cos B = \dfrac{a^2 + c^2 - b^2}{2\,ac}$
$ = \dfrac{10^2 + 15^2 - 20^2}{2 \cdot 10 \cdot 15}$
$\cos B = -0.25$
$m\angle B = 104.5$
$m\angle BAD = 180 - 104.5 = 75.5$
(Note: The angle between the two applied forces is $\angle BAD$, not $\angle B$.)

Algebra 2 and Trigonometry Review

PART 3. Probability and Statistics

VI. PROBABILITY

1. Probability and Counting Principle

Sample Space: all possible outcomes
Event: the favorable outcomes

(1) The probability of a simple event

$$P(E) = \frac{\text{number of the outcomes of the event}}{\text{number of the outcomes of the sample space}}$$

$$P(E) = \frac{n(E)}{n(S)}$$

e.g. A bag contains 6 black balls and 4 white balls. What is the probability of selecting a black ball?

$$P(\text{Black}) = \frac{n(\text{Black})}{n(\text{Sample Space})} = \frac{6}{10}$$

Impossible Case: $P(E) = 0$
Certain Case: $P(E) = 1$
Negation: $P(\text{Not } E) = 1 - P(E)$

e.g. If $P(\text{rain}) = 30\%$,
then $P(\text{Not rain}) = 1 - P(\text{rain}) = 70\%$

(2) Counting Principle (2 or more activities)

If the first activity can occur in M ways and the second activity can occur in N ways, then both activities can occur in M•N ways.

e.g. 3 doors to a building, 2 stairways to the second floor. There are $3 \cdot 2 = 6$ different ways to go.

Counting Principle for Probability

When A and B are indepenent events, the compound event of A and B has the probability
$P(A, B) = P(A) \cdot P(B)$

e.g. 4 students. The probability of the tallest one in the first place (A) and the shortest one in the last place (B)

$$P(A, B) = P(A) \cdot P(B) = \frac{1}{4} \cdot \frac{1}{3} = \frac{1}{12}$$

2. Permutation and Combination

In a **permutation** the order of the objects is important.
(1) The permutaion of n objects taken n at a time
$$_nP_n = n! = n(n-1)(n-2) \ldots 2 \cdot 1$$

e.g. Five letters A, B, C, D, E have 5! different arrangements. ($5! = 5 \cdot 4 \cdot 3 \cdot 2 \cdot 1 = 120$)

(2) The permutation of n objects taken n at a time with r items identical: $\dfrac{n!}{r!}$

e.g. Five letters COLOR have $\dfrac{5!}{2!}$ different arrangements.

e.g. Seven letters FREEZER, three E's identical and two R's identical, have $\dfrac{7!}{3! \cdot 2!}$ different arrangements.

(3) The permutation of n objects taken r ($r < n$) at a time
$$_nP_r = n(n-1)(n-2)\ldots \quad (r \text{ factors})$$

e.g. How many different arrangements of 1st, 2nd, and 3rd place are possible for 10 students?
$$_{10}P_3 = 10 \cdot 9 \cdot 8 = 720 \quad (3 \text{ factors})$$

In a **combination** the order of the objects does not matter.

e.g. (A, B, C) and (C, B, A) are considered same.

(4) The combination of n objects taken r at a time
$$_nC_r = \frac{_nP_r}{r!} \quad (r \leq n)$$

$_nC_n = 1$, $_nC_0 = 1$, $_nC_1 = n$, $_nC_r = {_nC_{n-r}}$

e.g. $_{50}C_{48} = {_{50}C_2}$ (to simplify the calculation)

e.g. How many 3 player teams can be formed from 10 students?
$$_{10}C_3 = \frac{_{10}P_3}{3!} = \frac{10 \cdot 9 \cdot 8}{3 \cdot 2 \cdot 1} = 120$$

e.g. From 10 boys and 12 girls, how many different teams can be formed if 2 members must be boys and 3 members must be girls?
$$_{10}C_2 \cdot {_{12}C_3} = \frac{10 \cdot 9}{2 \cdot 1} \cdot \frac{12 \cdot 11 \cdot 10}{3 \cdot 2 \cdot 1} = 9900$$

Using Graphing Calculator

e.g. $\quad 5! = 120$
$\quad$ [5] [MATH] PRB / 4: ! [ENTER]
e.g. $\quad {}_5P_3 = 60$
$\quad$ [5] [MATH] PRB / 2: $_nP_r$ [ENTER] 3 [ENTER]
e.g $\quad {}_7C_4 = 35$
$\quad$ [7] [MATH] PRB / 3: $_nC_r$ [ENTER] 4 [ENTER]

3. Binomial Probability (Bernoulli Experiment)

If the probability of success is p and the probability of failure is q = 1 - p, then the probability of exactly r successes in n independent trials is
$$_nC_r \, p^r \, q^{n-r}$$

e.g. The probability of rain on any given day is 0.3.
(1) The probability of rain on exactly 2 of 7 days is
$\quad P(2) = {}_7C_2(0.3)^2(0.7)^5 \quad$ (q = 1 - 0.3 = 0.7)
(2) The probability of rain at most 2 of 7 days is
$\quad$ P(at most 2) = P(0) + P(1) + P(2)
"at most 2 days" is same as "no more than 2 days"
(3) The probability of rain at least 4 of 7 days is
$\quad$ P(at least 4) = P(4) + P(5) + P(6) + P(7)
"at least 4 days" is same as "no less than 4 days"
(4) $\quad$ P(0) + P(1) + P(2) + • • • + P(6) + P(7) = 1
e.g. P(0) + P(1) + P(2) + • • • + P(6) = 1 - P(7)

Binomial Expansions:

$(x + y)^n = {}_nC_0 x^n y^0 + {}_nC_1 x^{n-1} y^1 + {}_nC_2 x^{n-2} y^2 \cdots {}_nC_n x^0 y^n$
There are n + 1 terms in the expansion.
The rth term is:
$\quad {}_nC_{(r-1)} x^{n-(r-1)} y^{(r-1)}$

e.g. $(x + y)^4 = 1 \cdot x^4 + 4 \cdot x^3 \cdot y + 6 \cdot x^2 \cdot y^2 + 4 \cdot x \cdot y^3 + 1 \cdot y^4$
e.g. $\quad (x + y)^8$
(1) the first term: $\quad$ (n = 8, r = 1, r - 1 = 0)
$\quad {}_8C_0 x^{8-0} y^0 = 1 \cdot x^8 \cdot 1 = x^8$
(2) the last term: $\quad$ (n = 8, r = 9, r - 1 = 8)
$\quad {}_8C_8 x^{8-8} y^8 = 1 \cdot x^0 \cdot y^8 = y^8$
(3) the middle term: $\quad$ (n = 8, r = 5, r - 1 = 4)
$\quad {}_8C_4 x^{8-4} y^4 = 70 x^4 y^4$
e.g. $\quad (2a - 1)^5$
The 3rd term is: $\quad$ (n = 5, r = 3, r - 1 = 2)
$\quad {}_5C_2 (2a)^{5-2} (-1)^2 = 10 \cdot (2a)^3 \cdot 1 = 80a^3$

VII. STATISTICS

1. Statistics (Univariate Data)

Common Methods of Collecting Data:

Surveys: get information through questionnaires, interviews, etc.
Controlled Experiments: consist of two groups of data, one of them served as a benchmark.
Observations: watch and study on the phenomena, without influence on the responses.

Size of the Data:

A **population** consists of the set of all items of interest.
A **sample** is a subset of items chosen from a population. The sample must be large enough to be effective and must be chosen **randomly** to eliminate any **bias**.

Analyze Data:

First arrange the data in numerical order.
Mean = Average = $\dfrac{\text{sum of the data values}}{\text{number of the data items}}$
Median: the middle value when the data arranged in order
Mode: the value that appears most often

Percentile: a number that tells what percent of the total number of the data values are less than or equal to a given data point.
1st Quartile (25th percentile): the middle value of the lower half set of the data, aka. **Lower Quartile**
2nd Quartile (50th percentile): the **median**, aka. **Middle Quartile**
3rd Quartile (75th percentile): the middle value of the upper half set of the data, aka. **Upper Quartile**

Range: the difference between the highest value and the lowest value.
Interquartile Range: the difference between the 3rd quartile value and the 1st quartile value.

Outliers: Some data points far outside most of the points in the data set.
Outliers can strongly affect the mean value. When outliers exist, use median to represent the central tendency of the data.

Algebra 2 and Trigonometry Review

e.g. Analyze the grades:

78, 85, 81, 95, 61, 85, 75, 88, 72, 100

First rearrange the data in numerical order:
61, 72, 75, 78, 81, 85, 85, 88, 95, 100
(make sure the number of items are same)

Mean = $\dfrac{820}{10}$ = 82

Median = $\dfrac{81 + 85}{2}$ = 83

(if the set has an even number of data values, take the average of the two middle values)

Mode = 85
Middle Quartile = Median = 83
Lower Quartile = 75
Upper Quartile = 88

Box-and-Whisker Plot :

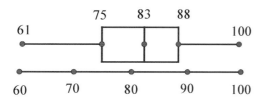

Range = 100 - 61, Interquartile Range = 88 - 75

Frequency Table :

Interval	Frequency
61 - 70	1
71 - 80	3
81 - 90	4
91 - 100	2

Frequency Histogram :

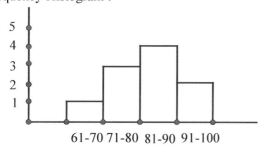

Variance for Populations:

$$v = \dfrac{1}{n} \sum_{i=1}^{n} (x_i - \bar{x})^2$$

Standard Deviation for Populations:
$\delta = \sqrt{v}$ (calculator symbol: δ_x)
n = number of the data items, $\bar{x}$ = mean, x_i = data values
Use graphing calculator to find $\bar{x}$ and δ_x.

Variance for Samples:

$$v = \dfrac{1}{n-1} \sum_{i=1}^{n} (x_i - \bar{x})^2$$

Standard Deviation for Samples:
$S = \sqrt{v}$ (calculator symbol: S_x)
Use graphing calculator to find $\bar{x}$ and S_x.

The difference between δ_x and S_x is $\dfrac{1}{n}$ and $\dfrac{1}{n-1}$.
With same set of data, S_x is slightly greater than δ_x.
This difference is insignificant when the sample is large enough.

Z-Score

$$\text{z-score} = \dfrac{x - \bar{x}}{\delta}$$

The z-score tells us how many standard deviations a data value x is above or below the mean.

Normal Distribution:
The properties of a normal curve:
(1). Symmetric ;
(2). Mean, Median, and Mode have the same value.
(3). 68.2% of the data values between $\bar{x} - \delta$ and $\bar{x} + \delta$.
  95.4% of the data values between $\bar{x} - 2\delta$ and $\bar{x} + 2\delta$.
  99.8% of the data values between $\bar{x} - 3\delta$ and $\bar{x} + 3\delta$.

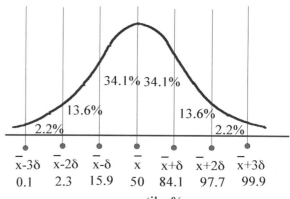

percentile %

Algebra 2 and Trigonometry Review

e.g. A normal distribution has a mean of 10.50 and a standard deviation of 0.75. What percent of the data values are in the range from 9.75 to 11.25 ?

$9.75 = 10.50 - 0.75 = \bar{x} - \delta$
$11.25 = 10.50 + 0.75 = \bar{x} + \delta$

There are 68.2% of the data values between 9.75 and 11.25.

e.g. The ages of the new teachers are normally distributed. 95.4% of the ages, centered about the mean, are between 24.6 and 37.4. Find the mean and the standard deviation.

$\bar{x} = \dfrac{24.6 + 37.4}{2} = 31$
$(\bar{x} + 2\delta) - (\bar{x} - 2\delta) = 4\delta = 37.4 - 24.6 = 12.8$
$\delta = 3.2$

Using Graphing Calculator

e.g. Find Mean and Standard Deviation
95, 92, 86, 84, 78

(1) Clear List L_1
[STAT] EDIT / 4: ClrList [ENTER] [2nd] [L_1]
[ENTER]
(2) Enter data to L_1
[STAT] EDIT / 1: Edit ... [ENTER]
    enter the above data into list L_1
(3) Display the One Variable Statistics
[STAT] CALC / 1: 1 - Var [ENTER] [2nd] [L_1]
[ENTER]
    $\bar{x} = 87$, $\delta x = 6$, $S_x = 6.7$
    $Q_1 = 81$, Med = 86, $Q_3 = 93.5$

e.g. x_i values, f_i frequency

x_i	92	87	82	77	72	67	62
f_i	2	3	6	9	10	6	4

(1) clear List L_1 and List L_2
[STAT] EDIT / 4: ClrList [ENTER] [2nd] [L_1] [,]
[2nd] [L_2] [ENTER]
(2) Enter data to L_1 and L_2
[STAT] EDIT / 1: Edit ... [ENTER]
    enter the x_i data into list L_1 and the f_i data into L_2
(3) Display the **One Variable** Statistics
[STAT] CALC / 1: 1 - Var [ENTER] [2nd] [L_1] [,]
[2nd] [L_2] [ENTER]
    $\bar{x} = 75$, $\delta x = 7.89$, $S_x = 7.99$,
    $Q_1 = 69.5$, Med = 74.5, $Q_3 = 82$
    $\bar{x} \neq$ Med   This is not a normal distribution.

2. Statistics (Bivariate Data)

Correlation: The relationship between two sets of data
Causation: The relationship in which one variable produces an effect on the other

Regression Modeling
Linear Regression: $y = ax + b$

Correlation Coefficient r :

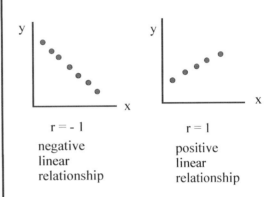

r = -1
negative linear relationship

r = 1
positive linear relationship

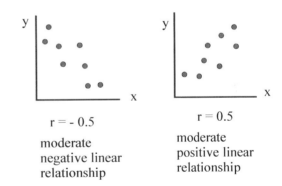

r = -0.5
moderate negative linear relationship

r = 0.5
moderate positive linear relationship

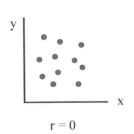

r = 0

no linear relationship

22. Algebra 2 and Trigonometry Review

Line of Best Fit (The Linear Regression)

The difference between the model values and the real values is the least.

Use graphing calculator to find the equation of the Line of Best Fit : $y = ax + b$

e.g.

x_i	2	4	6	8	10
y_i	13	15	16	17	20

(1) Clear List L_1 and List L_2:
[STAT] EDIT / 4: ClrList [ENTER]
[2nd] [L_1] [,] [2nd] [L_2] [ENTER]
(2) Enter data to L_1 and L_2:
[STAT] EDIT / 1: Edit ... [ENTER]
Enter data x_i into List L_1; Enter data y_i into List L_2.
(3) Scatter Plot:
[2nd] [STAT PLOT] 1: PLOT 1 [ENTER]

ON
Type: ⋰

[ZOOM] [9]

(4) Find the equation of the Line of Best Fit
     and the Correlation Coefficient r :
[2nd] [CATALOG] Diagnostic On [ENTER]
[STAT] CALC / 4: LinReg(ax + b) [ENTER]
[2nd] [L_1] [,] [L_2] [ENTER]
LinReg $y = ax + b$
       $a = 0.8$ $b = 11.4$ $r = 0.98$

(5) Draw the Line of Best Fit:
[Y =] [VARS] 5: Statistics ... [ENTER] EQ / 1: RegEQ [ENTER] [ZOOM] [9]

(6) Predict the results by using the model:
Find the value of y when x = 10.5
[2nd] [CALC] 1: Values [ENTER]
X = 10.5 [ENTER]   Y = 19.8
(Adjust Window Dimensions for Extrapolation)

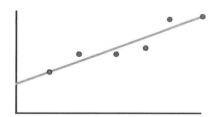

Other Regressions:

5: QuadReg: $y = ax^2 + bx + c$
6: CubicReg: $y = ax^3 + bx^2 + cx + d$
9: LnReg: $y = a + b\ln x$
0: ExpReg: $y = a \cdot b^x$
A: PwrReg: $y = ax^b$

To find the best model for a given set of data, compare the values of r.
Correlation Coefficient $r = \pm 1$ means exactly fit.

e.g.

x_i	0	1	2	3	4	5	6
y_i	5	10	20	40	80	160	320

(1) Enter the data into L_1 and L_2
(2) Make the scatter plot for the data
(3) Test the exponential model: $y = a \cdot b^x$
[STAT] CALC / 0: ExpReg [ENTER] [2nd] [L_1]
[,] [2nd] [L_2] [ENTER]
ExpReg $y = a \cdot b^x$ $a = 5$ $b = 2$ $r = 1$
         $y = 5 \cdot 2^x$
(4) Draw the graph of the model:
[Y =] [VARS] 5: Statistics ... [ENTER] EQ /
1: RegEQ [ENTER] [ZOOM] [9]

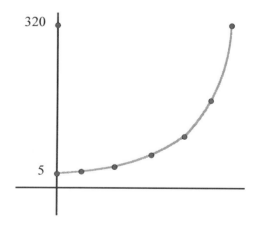

Algebra 2 and Trigonometry Review

PART 4. Graphing Calculator

VIII. GRAPHING CALCULATOR

1. Tips

Clear the Memory:
[2nd] [MEM] 7: Reset ... [ENTER]
1: All Ram ...[ENTER]   2: Reset [ENTER]
           Ram Cleared
[2nd] [MEM] 7: Reset ... [ENTER]
2: Defaults ...[ENTER]   2: Reset [ENTER]
           Defaults Set

Return to Home Screen:
[2nd] [QUIT]

2. Graph Functions

e.g. Graph $y = -x^2 + 4$
   [y =] [(-)] [X, T, θ, n] [x^2] [+] [4] [GRAPH]

e.g. Graph $y = |x - 4|$
   [y =] [MATH] NUM / 1: abs [ENTER]
     [X, T, θ, n] [-] [4] [ )] [GRAPH]

Zoom Menu:
To have a better view of the graph:
[ZOOM] 6: ZStandard (- 10 < x < 10 ; - 10 < y < 10)
[ZOOM] 4: ZDecimal   (to see friendly windows)
[ZOOM] 0: ZoomFit   (to see complete graphs)
[ZOOM] 2: Zoom In   (to see details around cursor)
[ZOOM] 7: ZTrig   (for Trigonometry ; $X_{scl} = \frac{\pi}{2}$)
[ZOOM] 9: ZoomStat   (for Statistics)
e.g. Graph $y = \sin 2x$
   [y =] [sin] [2] [X, T, θ, n] [ZOOM] [7]

Change Window Dimensions:
e.g. Graph $y = -3x^2 + 12x + 5$
[y =] [(-)] [3] [X, T, θ, n] [x^2] [+] [1] [2]
[X, T, θ, n] [+] [5] [ZOOM] [6]
  To see the complete graph:
    [WINDOW] Y_{max} = 20 [ENTER] [GRAPH]

Trace:
To see the y values vary with x values:
    [TRACE]
To find the value of y at a specific value of x:
    [2nd] [CALC] 1: value [ENTER]

3. Table of a Function

e.g. Display the table of $y = x^2$
   [y =] [X, T, θ, n] [x^2]
   [2nd] [TABLE]
To change the x increment:
   [2nd] [TBLSET] ΔTbl

4. Calculations

Solve Equations:

e.g. Solve $x^2 - 9 = 0$
 (1) graph $y = x^2 - 9$
 (2) [2nd] [CALC] 2: zero [ENTER]
 (3) move the cursor to set the Left Bound [ENTER]
    and the Right Bound [ENTER] of the x - intercept,
    then Guess [ENTER]
       Zero x = - 3   y = 0
 (4) repeat (3) to find the other zero
       Zero x = 3   y = 0

Solve the System of Equations:

e.g. solve system $xy = 8$
                $y = x + 2$
 (1) rewrite the first Eq. as $y = \frac{8}{x}$
 (2) graph those two functions
 (3) [2nd] [CALC] 5: intersect [ENTER]
 (4) $Y_1 = 8/X$
    First Curve?
    move the cursor to the intersection [ENTER]
    $Y_2 = X + 2$
    Second Curve ? [ENTER]
    [GUESS] [ENTER]
    Intersection X = 2   Y = 4
 (5) repeat (4) to find the other intersection:
      X = -4   Y = -2

Maximum and Minimum:

e.g. Find the maximum or minimum of the function
      $y = x^2 - 6x + 3$
 (1) graph $y = x^2 - 6x + 3$
 (2) [2nd] [CALC] 3: minimum [ENTER]
 (3) move the cursor to set the Left Bound [ENTER]
    and the Right Bound [ENTER] of the minimum,
    then Guess [ENTER]
       Minimum x = 3   y = -6
  when x = 3 the function has a minimum of -6 .

Algebra 2/Trigonometry Reference Sheet

Area of a Triangle

$K = \dfrac{1}{2} ab \sin C$

Functions of the Sum of Two Angles

$\sin(A + B) = \sin A \cos B + \cos A \sin B$

$\cos(A + B) = \cos A \cos B - \sin A \sin B$

$\tan(A + B) = \dfrac{\tan A + \tan B}{1 - \tan A \tan B}$

Functions of the Difference of Two Angles

$\sin(A - B) = \sin A \cos B - \cos A \sin B$

$\cos(A - B) = \cos A \cos B + \sin A \sin B$

$\tan(A - B) = \dfrac{\tan A - \tan B}{1 + \tan A \tan B}$

Law of Sines

$\dfrac{a}{\sin A} = \dfrac{b}{\sin B} = \dfrac{c}{\sin C}$

Sum of a Finite Arithmetic Series

$S_n = \dfrac{n(a_1 + a_n)}{2}$

Binomial Theorem

$(a + b)^n = {}_nC_0 a^n b^0 + {}_nC_1 a^{n-1} b^1 + {}_nC_2 a^{n-2} b^2 + \ldots + {}_nC_n a^0 b^n$

$(a + b)^n = \sum_{r=0}^{n} {}_nC_r a^{n-r} b^r$

Law of Cosines

$a^2 = b^2 + c^2 - 2bc \cos A$

Functions of the Double Angle

$\sin 2A = 2 \sin A \cos A$

$\cos 2A = \cos^2 A - \sin^2 A$

$\cos 2A = 2 \cos^2 A - 1$

$\cos 2A = 1 - 2 \sin^2 A$

$\tan 2A = \dfrac{2 \tan A}{1 - \tan^2 A}$

Functions of the Half Angle

$\sin \dfrac{1}{2} A = \pm \sqrt{\dfrac{1 - \cos A}{2}}$

$\cos \dfrac{1}{2} A = \pm \sqrt{\dfrac{1 + \cos A}{2}}$

$\tan \dfrac{1}{2} A = \pm \sqrt{\dfrac{1 - \cos A}{1 + \cos A}}$

Sum of a Finite Geometric Series

$S_n = \dfrac{a_1(1 - r^n)}{1 - r}$

Normal Curve Standard Deviation

0.1% 0.5% 1.7% 4.4% 9.2% 15.0% 19.1% 19.1% 15.0% 9.2% 4.4% 1.7% 0.5% 0.1%

−3 −2.5 −2 −1.5 −1 −0.5 0 0.5 1 1.5 2 2.5 3

Practice   **I. Algebraic Expressions, Equations, and Inequalities**

1. Factoring Polynomials

1. Factored completely, the expression $6x - x^3 - x^2$ is equivalent to
(1) $x(x+3)(x-2)$
(2) $x(x-3)(x+2)$
(3) $-x(x-3)(x+2)$
(4) $-x(x+3)(x-2)$

2. Factored completely, the expression $12x^4 + 10x^3 - 12x^2$ is equivalent to
(1) $x^2(4x+6)(3x-2)$
(2) $2(2x^2+3x)(3x^2-2x)$
(3) $2x^2(2x-3)(3x+2)$
(4) $2x^2(2x+3)(3x-2)$

2. Quadratic Equations and Inequalities

3. If $x^2 + 2 = 6x$ is solved by completing the square, an intermediate step would be
(1) $(x+3)^2 = 7$
(2) $(x-3)^2 = 7$
(3) $(x-3)^2 = 11$
(4) $(x-6)^2 = 34$

4. The solutions of the equation $y^2 - 3y = 9$ are
(1) $\dfrac{3 \pm 3i\sqrt{3}}{2}$
(2) $\dfrac{3 \pm 3i\sqrt{5}}{2}$
(3) $\dfrac{-3 \pm 3\sqrt{5}}{2}$
(4) $\dfrac{3 \pm 3\sqrt{5}}{2}$

5. The roots of the equation $2x^2 + 7x - 3 = 0$ are
(1) $-\dfrac{1}{2}$ and -3
(2) $\dfrac{1}{2}$ and 3
(3) $\dfrac{-7 \pm \sqrt{73}}{4}$
(4) $\dfrac{7 \pm \sqrt{73}}{4}$

6. The solution set of the inequality $x^2 - 3x > 10$ is
(1) $\{x \mid -2 < x < 5\}$
(2) $\{x \mid 0 < x < 3\}$
(3) $\{x \mid x < -2 \text{ or } x > 5\}$
(4) $\{x \mid x < -5 \text{ or } x > 2\}$

7. Which values of x are in the solution set of the following system of equations?
$$y = 3x - 6$$
$$y = x^2 - x - 6$$
(1) $0, -4$ (2) $0, 4$ (3) $6, -2$ (4) $-6, 2$

I. Algebraic Expressions, Equations, and Inequalities

3. Rational Expressions and Operations

8. The expression $\dfrac{a^2 b^{-3}}{a^{-4} b^2}$ is equivalent to

   (1) $\dfrac{a^6}{b^5}$ (2) $\dfrac{b^5}{a^6}$ (3) $\dfrac{a^2}{b}$ (4) $a^{-2} b^{-1}$

9. Written in simplest form, the expression $\dfrac{\dfrac{x}{4} - \dfrac{1}{x}}{\dfrac{1}{2x} + \dfrac{1}{4}}$ is equivalent to

   (1) $x - 1$ (2) $x - 2$ (3) $\dfrac{x-2}{2}$ (4) $\dfrac{x^2-4}{x+2}$

10. When $x^{-1} - 1$ is divided by $x - 1$, the quotient is

    (1) -1 (2) $-\dfrac{1}{x}$ (3) $\dfrac{1}{x^2}$ (4) $\dfrac{1}{(x-1)^2}$

11. When $\dfrac{3}{2}x^2 - \dfrac{1}{4}x - 4$ is subtracted from $\dfrac{5}{2}x^2 - \dfrac{3}{4}x + 1$, the difference is

    (1) $-x^2 + \dfrac{1}{2}x - 5$ (2) $x^2 - \dfrac{1}{2}x + 5$ (3) $-x^2 - x - 3$ (4) $x^2 - x - 3$

4. Radicals

12. The expression $4ab\sqrt{2b} - 3a\sqrt{18b^3} + 7ab\sqrt{6b}$ is equivalent to
    (1) $2ab\sqrt{6b}$
    (2) $16ab\sqrt{2b}$
    (3) $-5ab + 7ab\sqrt{6b}$
    (4) $-5ab\sqrt{2b} + 7ab\sqrt{6b}$

13. The product of $(3 + \sqrt{5})$ and $(3 - \sqrt{5})$ is
    (1) $4 - 6\sqrt{5}$ (2) $14 - 6\sqrt{5}$ (3) 14 (4) 4

14. If $a = 3$ and $b = -2$, what is the value of the expression $\dfrac{a^{-2}}{b^{-3}}$?

    (1) $-\dfrac{9}{8}$ (2) -1 (3) $-\dfrac{8}{9}$ (4) $\dfrac{8}{9}$

15. The expression $(x^2 - 1)^{-\frac{2}{3}}$ is equivalent to

    (1) $\sqrt[3]{(x^2-1)^2}$ (2) $\dfrac{1}{\sqrt[3]{(x^2-1)^2}}$ (3) $\sqrt{(x^2-1)^3}$ (4) $\dfrac{1}{\sqrt{(x^2-1)^3}}$

Practice   I. Algebraic Expressions, Equations, and Inequalities   27.

16. When simplified, the expression $\left(\dfrac{w^{-5}}{w^{-9}}\right)^{\frac{1}{2}}$ is equivalent to

(1) w^{-7} (2) w^2 (3) w^7 (4) w^{14}

17. The expression $x^{-\frac{2}{5}}$ is equivalent to

(1) $-\sqrt[2]{x^5}$ (2) $-\sqrt[5]{x^2}$ (3) $\dfrac{1}{\sqrt[2]{x^5}}$ (4) $\dfrac{1}{\sqrt[5]{x^2}}$

18. Which expression is equivalent to $\dfrac{\sqrt{3}+5}{\sqrt{3}-5}$?

(1) $-\dfrac{14+5\sqrt{3}}{11}$ (2) $-\dfrac{17+5\sqrt{3}}{11}$ (3) $\dfrac{14+5\sqrt{3}}{14}$ (4) $\dfrac{17+5\sqrt{3}}{14}$

19. The fraction $\dfrac{3}{\sqrt{3a^2b}}$ is equivalent to

(1) $\dfrac{1}{a\sqrt{b}}$ (2) $\dfrac{\sqrt{b}}{ab}$ (3) $\dfrac{\sqrt{3b}}{ab}$ (4) $\dfrac{\sqrt{3}}{a}$

20. The expression $\dfrac{2x+4}{\sqrt{x+2}}$ is equivalent to

(1) $\dfrac{(2x+4)\sqrt{x-2}}{x-2}$ (2) $\dfrac{(2x+4)\sqrt{x-2}}{x-4}$ (3) $2\sqrt{x-2}$ (4) $2\sqrt{x+2}$

21. The solution set of the equation $\sqrt{x+3} = 3-x$ is
(1) {1}   (2) {0}   (3) {1, 6}   (4) {2, 3}

5. Absolute Value Equations and Inequalities

22. Which graph represents the solution set of $|6x-7| \leq 5$?

(1) [number line with closed interval from $\frac{1}{3}$ to 2]   (3) [number line shaded left of $-\frac{1}{3}$]

(2) [number line shaded outside $\frac{1}{3}$ and 2]   (4) [number line shaded right of $-\frac{1}{3}$]

23. What is the solution set of the equation $|4a+6| - 4a = -10$?

(1) ∅   (2) {0}   (3) $\left\{\dfrac{1}{2}\right\}$ (4) $\left\{0, \dfrac{1}{2}\right\}$

I. Algebraic Expressions, Equations, and Inequalities

6. Complex Numbers

24. In simplest form, $\sqrt{-300}$ is equivalent to
(1) $3i\sqrt{10}$
(2) $5i\sqrt{12}$
(3) $10i\sqrt{3}$
(4) $12i\sqrt{5}$

25. The product of i^7 and i^5 is equivalent to
(1) 1
(2) -1
(3) i
(4) $-i$

26. The expression $2i^2 + 3i^3$ is equivalent to
(1) $-2-3i$
(2) $2-3i$
(3) $-2+3i$
(4) $2+3i$

27. The expression $(3-7i)^2$ is equivalent to
(1) $-40+0i$
(2) $-40-42i$
(3) $58+0i$
(4) $58-42i$

28. What is the conjugate of $-2+3i$?
(1) $-3+2i$
(2) $-2-3i$
(3) $2-3i$
(4) $3+2i$

29. The conjugate of $7-5i$ is
(1) $-7-5i$
(2) $-7+5i$
(3) $7-5i$
(4) $7+5i$

7. Roots of the Quadratic Equations

30. For which equation does the sum of the roots equal $\frac{3}{4}$ and the product of the roots equal -2?
(1) $4x^2 - 8x + 3 = 0$
(2) $4x^2 + 8x + 3 = 0$
(3) $4x^2 - 3x - 8 = 0$
(4) $4x^2 + 3x - 2 = 0$

31. The roots of the equation $9x^2 + 3x - 4 = 0$ are
(1) imaginary
(2) real, rational, and equal
(3) real, rational, and unequal
(4) real, irrational, and unequal

32. The roots of the equation $x^2 - 10x + 25 = 0$ are
(1) imaginary
(2) real and irrational
(3) real, rational, and equal
(4) real, rational, and unequal

33. For which equation does the sum of the roots equal -3 and the product of the roots equal 2?
(1) $x^2 + 2x - 3 = 0$
(2) $x^2 - 3x + 2 = 0$
(3) $2x^2 + 6x + 4 = 0$
(4) $2x^2 - 6x + 4 = 0$

Practice   **I. Algebraic Expressions, Equations, and Inequalities**   **29.**

8. Polynomial Equations of Higher Degrees

34. The graph of $y = f(x)$ is shown below.

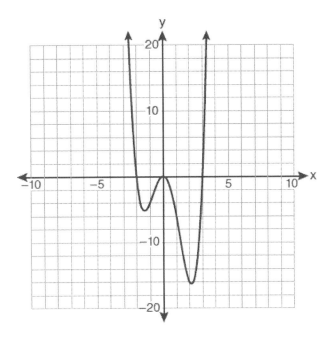

Which set lists all the real solutions of $f(x) = 0$?
(1) $\{-3, 2\}$ (2) $\{-2, 3\}$ (3) $\{-3, 0, 2\}$ (4) $\{-2, 0, 3\}$

35. The graph of $y = x^3 - 4x^2 + x + 6$ is shown below.

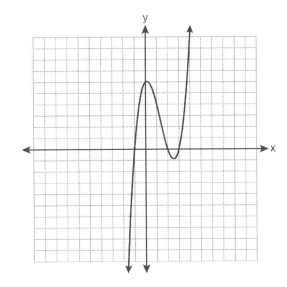

What is the product of the roots of the equation $x^3 - 4x^2 + x + 6 = 0$?
(1) -36 (2) -6 (3) 6 (4) 4

30. **I. Algebraic Expressions, Equations, and Inequalities**

36. Which values of x are solutions of the equation $x^3 + x^2 - 2x = 0$?
 (1) 0, 1, 2 (3) 0, −1, 2
 (2) 0, 1, −2 (4) 0, −1, −2

9. Sequence and Series

37. The value of the expression $2\sum_{n=0}^{2}(n^2 + 2^n)$ is
 (1) 12 (2) 22 (3) 24 (4) 26

38. Mrs. Hill asked her students to express the sum $1 + 3 + 5 + 7 + 9 + \ldots + 39$ using sigma notation. Four different student answers were given. Which student answer is correct?
 (1) $\sum_{k=1}^{20}(2k-1)$ (2) $\sum_{k=2}^{40}(k-1)$ (3) $\sum_{k=-1}^{37}(k+2)$ (4) $\sum_{k=1}^{39}(2k-1)$

39. What is the common difference of the arithmetic sequence $5, 8, 11, 14$?
 (1) $\frac{8}{5}$ (2) −3 (3) 3 (4) 9

40. What is a formula for the nth term of sequence B shown below?
 $B = 10, 12, 14, 16, \ldots$
 (1) $b_n = 8 + 2n$ (2) $b_n = 10 + 2n$ (3) $b_n = 10(2)^n$ (4) $b_n = 10(2)^{n-1}$

41. Which arithmetic sequence has a common difference of 4?
 (1) $\{0, 4n, 8n, 12n, \ldots\}$ (3) $\{n+1, n+5, n+9, n+13, \ldots\}$
 (2) $\{n, 4n, 16n, 64n, \ldots\}$ (4) $\{n+4, n+16, n+64, n+256, \ldots\}$

42. What is the formula for the nth term of the sequence $54, 18, 6, \ldots$?
 (1) $a_n = 6\left(\frac{1}{3}\right)^n$ (2) $a_n = 6\left(\frac{1}{3}\right)^{n-1}$ (3) $a_n = 54\left(\frac{1}{3}\right)^n$ (4) $a_n = 54\left(\frac{1}{3}\right)^{n-1}$

43. What is the common ratio of the geometric sequence whose first term is 27 and fourth term is 64?
 (1) $\frac{3}{4}$ (2) $\frac{64}{81}$ (3) $\frac{4}{3}$ (4) $\frac{37}{3}$

44. What is the fifteenth term of the sequence $5, -10, 20, -40, 80, \ldots$?
 (1) −163,840 (2) −81,920 (3) 81,920 (4) 327,680

Practice I. Algebraic Expressions, Equations, and Inequalities

Show Work:

Polynomials, Rationals, Radicals, Absolute Values

1. Factor completely: $10ax^2 - 23ax - 5a$

2. Solve $2x^2 - 12x + 4 = 0$ by completing the square, expressing the result in simplest radical form.

3. Express in simplest form: $\dfrac{\dfrac{1}{2} - \dfrac{4}{d}}{\dfrac{1}{d} + \dfrac{3}{2d}}$

4. Solve for x: $\dfrac{4x}{x-3} = 2 + \dfrac{12}{x-3}$

5. Solve algebraically for x: $\dfrac{1}{x+3} - \dfrac{2}{3-x} = \dfrac{4}{x^2-9}$

6. Express $5\sqrt{3x^3} - 2\sqrt{27x^3}$ in simplest radical form.

7. Express $\left(\dfrac{2}{3}x - 1\right)^2$ as a trinomial.

8. Express $\dfrac{5}{3-\sqrt{2}}$ with a rational denominator, in simplest radical form.

9. Express $\dfrac{\sqrt{108x^5y^8}}{\sqrt{6xy^5}}$ in simplest radical form.

Practice I. Algebraic Expressions, Equations, and Inequalities 33.

Complex Numbers, Polynomial Equations, Roots, Sequence and Series

10. Use the discriminant to determine all value of k that would result in the equation $x^2 - kx + 4 = 0$ having equal roots.

11. Find the sum and product of the roots of the equation $5x^2 + 11x - 3 = 0$.

12. Solve the equation $8x^3 + 4x^2 - 18x - 9 = 0$ algebraically for all values of x.

13. Find the first four terms of the recursive sequence defined below.
$$a_1 = -3$$
$$a_n = a_{(n-1)} - n$$

14. Express the sum $7 + 14 + 21 + 28 + \ldots + 105$ using sigma notation.

15. Evaluate: $10 + \sum_{n=1}^{5} (n^3 - 1)$

II. Relations and Functions

1. Which function is *not* one-to-one?
   (1) $\{(0,1),(1,2),(2,3),(3,4)\}$
   (2) $\{(0,0),(1,1),(2,2),(3,3)\}$
   (3) $\{(0,1),(1,0),(2,3),(3,2)\}$
   (4) $\{(0,1),(1,0),(2,0),(3,2)\}$

2. What is the domain of the function $f(x) = \sqrt{x-2} + 3$?
   (1) $(-\infty, \infty)$
   (2) $(2, \infty)$
   (3) $[2, \infty)$
   (4) $[3, \infty)$

3. What are the domain and the range of the function shown in the graph below?

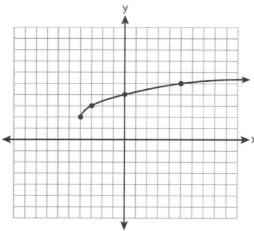

   (1) $\{x \mid x > -4\}; \{y \mid y > 2\}$
   (2) $\{x \mid x \geq -4\}; \{y \mid y \geq 2\}$
   (3) $\{x \mid x > 2\}; \{y \mid y > -4\}$
   (4) $\{x \mid x \geq 2\}; \{y \mid y \geq -4\}$

4. Which graph does *not* represent a function?
   (1)   (2)   (3)   (4)

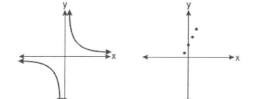

5. The equation $y - 2\sin\theta = 3$ may be rewritten as
   (1) $f(y) = 2\sin x + 3$
   (2) $f(y) = 2\sin\theta + 3$
   (3) $f(x) = 2\sin\theta + 3$
   (4) $f(\theta) = 2\sin\theta + 3$

6. Which relation is *not* a function?
   (1) $(x-2)^2 + y^2 = 4$
   (2) $x^2 + 4x + y = 4$
   (3) $x + y = 4$
   (4) $xy = 4$

Practice   **II. Relations and Functions**   **35.**

7. Which graph represents a one-to-one function?

(1)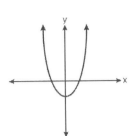

(3)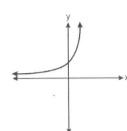

(2)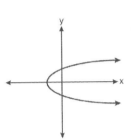

(4)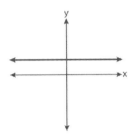

8. Which graph does *not* represent a function?

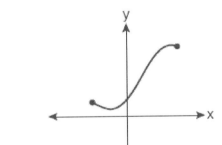

(1)   (3)

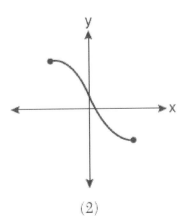 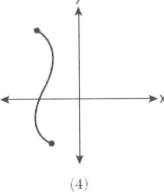

(2)   (4)

36. **II. Relations and Functions**

9. If $f(x) = \frac{1}{2}x - 3$ and $g(x) = 2x + 5$, what is the value of $(g \circ f)(4)$?
(1)   −13   (2)   3.5   (3)   3   (4)   6

10. If $f(x) = x^2 - 5$ and $g(x) = 6x$, then $g(f(x))$ is equal to
(1) $6x^3 - 30x$ (3) $36x^2 - 5$
(2) $6x^2 - 30$ (4) $x^2 + 6x - 5$

11. The minimum point on the graph of the equation $y = f(x)$ is $(-1, -3)$. What is the minimum point on the graph of the equation $y = f(x) + 5$?
(1) $(-1, 2)$ (2) $(-1, -8)$ (3) $(4, -3)$ (4) $(-6, -3)$

12. The graph below shows the function $f(x)$.

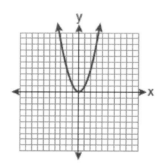

Which graph represents the function $f(x + 2)$?
(1)
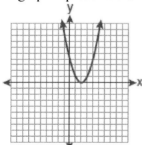

(3)
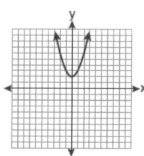

(2)
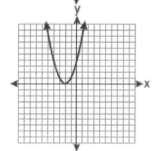

(4)
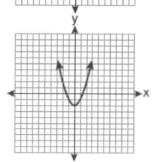

Practice II. Relations and Functions

Important Functions and Relations

13. Which graph best represents the inequality $y + 6 \geq x^2 - x$?

(1)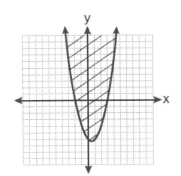

(3)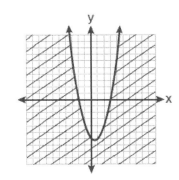

(2)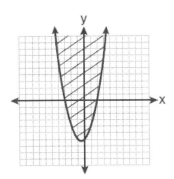

(4)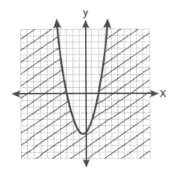

14. The equation $x^2 + y^2 - 2x + 6y + 3 = 0$ is equivalent to

(1) $(x - 1)^2 + (y + 3)^2 = -3$
(2) $(x - 1)^2 + (y + 3)^2 = 7$
(3) $(x + 1)^2 + (y + 3)^2 = 7$
(4) $(x + 1)^2 + (y + 3)^2 = 10$

15. Which equation represents the circle shown in the graph below that passes through the point $(0,-1)$?

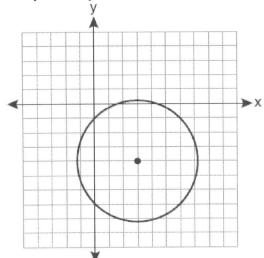

(1) $(x - 3)^2 + (y + 4)^2 = 16$
(2) $(x - 3)^2 + (y + 4)^2 = 18$
(3) $(x + 3)^2 + (y - 4)^2 = 16$
(4) $(x + 3)^2 + (y - 4)^2 = 18$

16. Four points on the graph of the function f(x) are shown below.
$$\{(0, 1), (1, 2), (2, 4), (3, 8)\}$$
Which equation represents f(x)?
(1) $f(x) = 2^x$ (2) $f(x) = 2x$ (3) $f(x) = x + 1$ (4) $f(x) = \log_2 x$

17. The solution set of $4^{x^2+4x} = 2^{-6}$ is
(1) $\{1, 3\}$ (2) $\{-1, 3\}$ (3) $\{-1, -3\}$ (4) $\{1, -3\}$

18. What is the value of x in the equation $9^{3x+1} = 27^{x+2}$?
(1)   1   (2) $\frac{1}{3}$ (3) $\frac{1}{2}$ (4) $\frac{4}{3}$

19. On January 1, a share of a certain stock cost $180. Each month thereafter, the cost of a share of this stock decreased by one-third. If x represents the time, in months, and y represents the cost of the stock, in dollars, which graph best represents the cost of a share over the following 5 months?

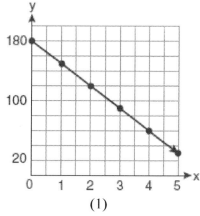
(1)

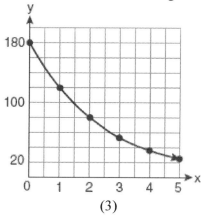
(3)

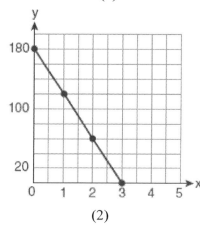
(2)

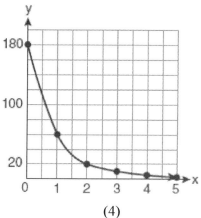
(4)

Practice   II. Relations and Functions   39.

20. The expression $\log_5\left(\dfrac{1}{25}\right)$ is equivalent to

(1) $\dfrac{1}{2}$ (2)   2   (3) $-\dfrac{1}{2}$ (4) -2

21. The expression $\log_8 64$ is equivalent to

(1)   8   (2)   2   (3) $\dfrac{1}{2}$ (4) $\dfrac{1}{8}$

22. The expression $2\log x - (3\log y + \log z)$ is equivalent to

(1) $\log\dfrac{x^2}{y^3 z}$ (2) $\log\dfrac{x^2 z}{y^3}$ (3) $\log\dfrac{2x}{3yz}$ (4) $\log\dfrac{2xz}{3y}$

23. What is the solution of the equation $2\log_4(5x) = 3$?

(1)   6.4   (2)   2.56   (3) $\dfrac{9}{5}$ (4) $\dfrac{8}{5}$

24. Which two functions are inverse functions of each other?

(1) $f(x) = \sin x$ and $g(x) = \cos(x)$ (3) $f(x) = e^x$ and $g(x) = \ln x$

(2) $f(x) = 3 + 8x$ and $g(x) = 3 - 8x$ (4) $f(x) = 2x - 4$ and $g(x) = -\dfrac{1}{2}x + 4$

25. If a function is defined by the equation $f(x) = 4^x$, which graph represents the inverse of this function?

(1)
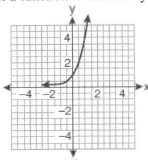

(3)
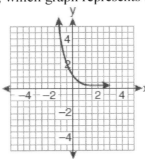

(2)
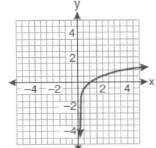

(4)
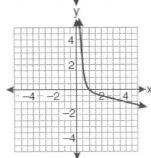

40.   **II. Relations and Functions**

Show Work:

1. The graph below represents the function $y = f(x)$.

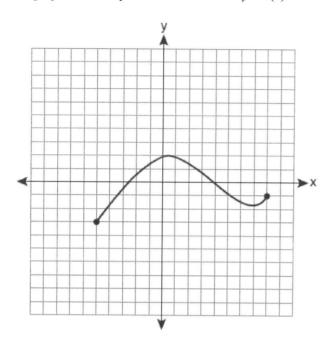

State the domain and range of this function.

2. For a given set of rectangles, the length is inversely proportional to the width. In one of these rectangles, the length is 12 and the width is 6. For this set of rectangles, calculate the width of a rectangle whose length is 9.

3. Write an equation of the circle shown in the graph below.

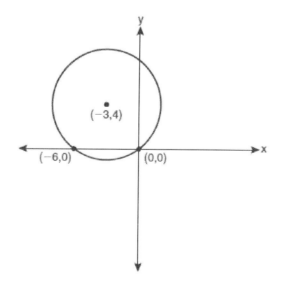

Practice   II. Relations and Functions   41.

4. A circle shown in the diagram below has a center of $(-5, 3)$ and passes through point $(-1, 7)$.

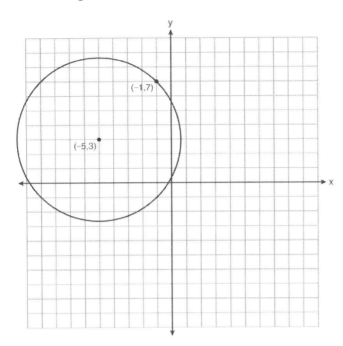

Write an equation that represents the circle.

5. Matt places $1,200 in an investment account earning an annual rate of 6.5%, compounded continuously. Using the formula $V = Pe^{rt}$, where V is the value of the account in t years, P is the principal initially invested, e is the base of a natural logarithm, and r is the rate of interest, determine the amount of money, to the *nearest cent*, that Matt will have in the account after 10 years.

6. The graph of the equation $y = \left(\dfrac{1}{2}\right)^x$ has an asymptote. On the grid below, sketch the graph of $y = \left(\dfrac{1}{2}\right)^x$ and write the equation of this asymptote.

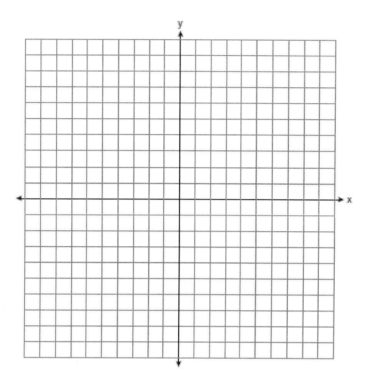

7. Solve algebraically for x: $16^{2x+3} = 64^{x+2}$

8. The temperature, T, of a given cup of hot chocolate after it has been cooling for t minutes can best be modeled by the function below, where T_0 is the temperature of the room and k is a constant.
$$\ln(T - T_0) = -kt + 4.718$$

A cup of hot chocolate is placed in a room that has a temperature of 68°. After 3 minutes, the temperature of the hot chocolate is 150°. Compute the value of k to the *nearest thousandth*. [Only an algebraic solution can receive full credit.]

Using this value of k, find the temperature, T, of this cup of hot chocolate if it has been sitting in this room for a total of 10 minutes. Express your answer to the *nearest degree*. [Only an algebraic solution can receive full credit.]

9. Solve algebraically for x: $\log_{x+3} \dfrac{x^3 + x - 2}{x} = 2$

44. **III. Trigonometric Functions**

1. What is the radian measure of an angle whose measure is −420°?
   (1) $-\dfrac{7\pi}{3}$ (2) $-\dfrac{7\pi}{6}$ (3) $\dfrac{7\pi}{6}$ (4) $\dfrac{7\pi}{3}$

2. In which graph is θ coterminal with an angle of −70°?
   (1)
   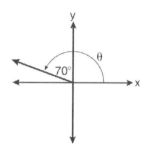
   (3)
   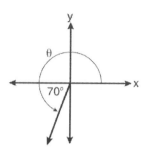
   (2)
   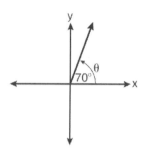
   (4)
   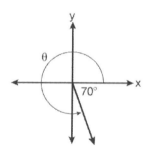

3. A circle has a radius of 4 inches. In inches, what is the length of the arc intercepted by a central angle of 2 radians?
   (1) 2π (2) 2   (3) 8π (4) 8

4. What is the number of degrees in an angle whose radian measure is $\dfrac{11\pi}{12}$?
   (1) 150   (2) 165   (3) 330   (4) 518

5. If $\angle A$ is acute and $\tan A = \dfrac{2}{3}$, then
   (1) $\cot A = \dfrac{2}{3}$ (2) $\cot A = \dfrac{1}{3}$ (3) $\cot(90° - A) = \dfrac{2}{3}$ (4) $\cot(90° - A) = \dfrac{1}{3}$

Practice   III. Trigonometric Functions   45.

6. In the diagram below of right triangle KTW, KW = 6, KT = 5, and m∠KTW = 90.

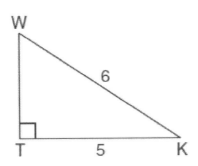

What is the measure of ∠K, to the *nearest minute*?
(1)   33°33'   (2)   33°34'   (3)   33°55'   (4)   33°56'

7. Which ratio represents csc A in the diagram below?

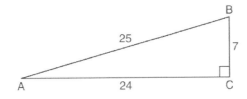

(1) $\dfrac{25}{24}$ (2) $\dfrac{25}{7}$ (3) $\dfrac{24}{7}$ (4) $\dfrac{7}{24}$

8. In the diagram below of right triangle JTM, JT = 12, JM = 6, and m∠JMT = 90.

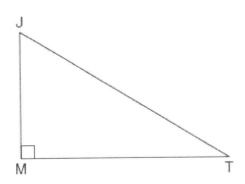

What is the value of cot J ?
(1) $\dfrac{\sqrt{3}}{3}$ (2) 2   (3) $\sqrt{3}$ (4) $\dfrac{2\sqrt{3}}{3}$

46.   **III. Trigonometric Functions**

9. In the diagram below of a unit circle, the ordered pair $(-\frac{\sqrt{2}}{2}, -\frac{\sqrt{2}}{2})$ represents the point where the terminal side of θ intersects the unit circle.

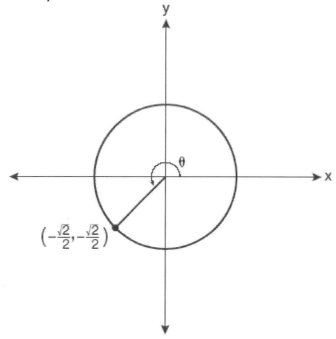

What is m∠θ?
(1)   45   (2)   135   (3)   225   (4)   240

Show Work:

1. Find, to the *nearest tenth of a degree*, the angle whose measure is 2.5 radians.

2. Find, to the *nearest minute*, the angle whose measure is 3.45 radians.

Practice III. Trigonometric Functions 47.

3. If θ is an angle in standard position and its terminal side passes through the point $(-3, 2)$, find the exact value of $\csc \theta$.

4. On the unit circle shown in the diagram below, sketch an angle, in standard position, whose degree measure is 240 and find the exact value of $\sin 240°$.

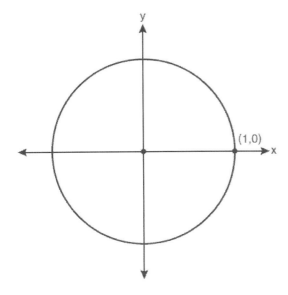

IV. Trigonometric Graphs

1. Which equation is represented by the graph below?

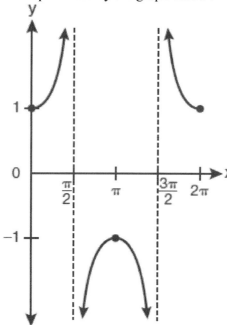

(1)　　$y = \cot x$　　(2)　　$y = \csc x$　　(3)　　$y = \sec x$　　(4)　　$y = \tan x$

2. Which graph represents the equation $y = \cos^{-1} x$?

(1)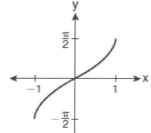

(3)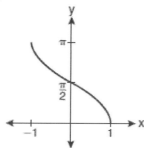

(2)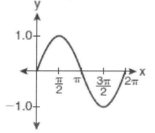

(4)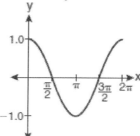

Practice IV. Trigonometric Graphs 49.

3. Which equation is sketched in the diagram below?

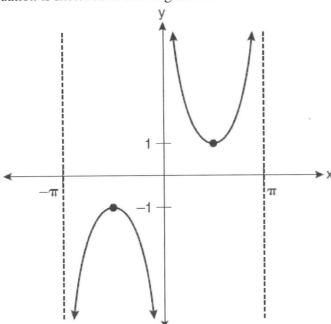

(1) $y = \csc x$ (2) $y = \sec x$ (3) $y = \cot x$ (4) $y = \tan x$

4. What is the period of the function $y = \frac{1}{2} \sin\left(\frac{x}{3} - \pi\right)$?

(1) $\frac{1}{2}$ (2) $\frac{1}{3}$ (3) $\frac{2}{3}\pi$ (4) 6π

5. Which graph represents one complete cycle of the equation $y = \sin 3\pi x$?

(1)
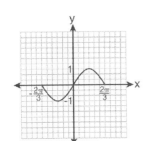

(3)
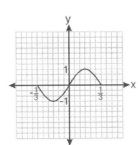

(2)
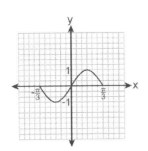

(4)
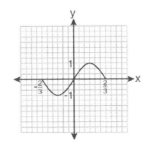

IV. Trigonometric Graphs

6. The function $f(x) = \tan x$ is defined in such a way that $f^{-1}(x)$ is a function. What can be the domain of $f(x)$?

(1) $\{x \mid 0 \leq x \leq \pi\}$ (2) $\{x \mid 0 \leq x \leq 2\pi\}$ (3) $\left\{x \mid -\dfrac{\pi}{2} < x < \dfrac{\pi}{2}\right\}$ (4) $\left\{x \mid -\dfrac{\pi}{2} < x < \dfrac{3\pi}{2}\right\}$

7. What is the principal value of $\cos^{-1}\left(-\dfrac{\sqrt{3}}{2}\right)$?

(1) $-30°$ (2) $60°$ (3) $150°$ (4) $240°$

8. If $\sin^{-1}\left(\dfrac{5}{8}\right) = A$, then

(1) $\sin A = \dfrac{5}{8}$ (2) $\sin A = \dfrac{8}{5}$ (3) $\cos A = \dfrac{5}{8}$ (4) $\cos A = \dfrac{8}{5}$

9. In physics class, Eva noticed the pattern shown in the accompanying diagram on an oscilloscope.

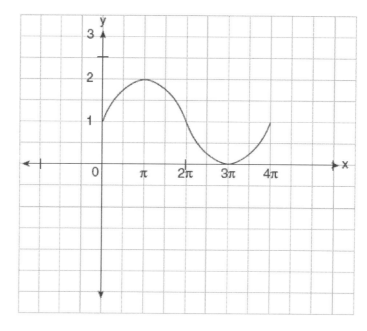

Which equation best represents the pattern shown on this oscilloscope?

(1) $y = \sin(\dfrac{1}{2}x) + 1$ (3) $y = 2\sin x + 1$

(2) $y = \sin x + 1$ (4) $y = 2\sin(-\dfrac{1}{2}x) + 1$

Practice   V. Trigonometric Applications

1. The expression $\cos 4x \cos 3x + \sin 4x \sin 3x$ is equivalent to

(1) $\sin x$ (2) $\sin 7x$ (3) $\cos x$ (4) $\cos 7x$

2. The expression $\cos^2 \theta - \cos 2\theta$ is equivalent to
(1) $\sin^2 \theta$ (2) $-\sin^2 \theta$ (3) $\cos^2 \theta + 1$ (4) $-\cos^2 \theta - 1$

3. If $\sin A = \dfrac{2}{3}$ where $0° < A < 90°$, what is the value of $\sin 2A$?

(1) $\dfrac{2\sqrt{5}}{3}$ (2) $\dfrac{2\sqrt{5}}{9}$ (3) $\dfrac{4\sqrt{5}}{9}$ (4) $-\dfrac{4\sqrt{5}}{9}$

4. What are the values of θ in the interval $0° \leq \theta < 360°$ that satisfy the equation $\tan \theta - \sqrt{3} = 0$?

(1) 60°, 240°   (3) 72°, 108°, 252°, 288°
(2) 72°, 252°   (4) 60°, 120°, 240°, 300°

5. In $\triangle ABC$, m$\angle A$ = 120, b = 10, and c = 18. What is the area of $\triangle ABC$ to the *nearest square inch*?
(1)   52   (2)   78   (3)   90   (4)   156

6. In $\triangle ABC$, a = 3, b = 5, and c = 7. What is m$\angle C$?
(1)   22   (2)   38   (3)   60   (4)   120

7. The sides of a parallelogram measure 10 cm and 18 cm. One angle of the parallelogram measures 46 degrees. What is the area of the parallelogram, to the *nearest square centimeter*?
(1)   65   (2)   125   (3)   129   (4)   162

8. In $\triangle ABC$, m$\angle A$ = 74, a = 59.2, and c = 60.3. What are the two possible values for m$\angle C$, to the *nearest tenth*?
(1) 73.7 and 106.3   (2) 73.7 and 163.7   (3) 78.3 and 101.7   (4) 78.3 and 168.3

9. How many distinct triangles can be formed if m$\angle A$ = 35, a = 10, and b = 13?
(1)   1   (2)   2   (3)   3   (4)   0

52.   V. Trigonometric Applications

Show Work:

1. If $\tan A = \dfrac{2}{3}$ and $\sin B = \dfrac{5}{\sqrt{41}}$ and angles A and B are in Quadrant I, find the value of $\tan(A+B)$.

2. Starting with $\sin^2 A + \cos^2 A = 1$, derive the formula $\tan^2 A + 1 = \sec^2 A$.

3. Find all values of θ in the interval $0° \leq \theta < 360°$ that satisfy the equation $\sin 2\theta = \sin \theta$.

4. Solve the equation $2\tan C - 3 = 3\tan C - 4$ algebraically for all values of C in the interval $0° \leq C < 360°$.

5. Two sides of a parallelogram are 24 feet and 30 feet. The measure of the angle between these sides is $57°$. Find the area of the parallelogram, to the *nearest square foot*.

Practice V. Trigonometric Applications

6. In a triangle, two sides that measure 6 cm and 10 cm form an angle that measures 80°. Find, to the *nearest degree*, the measure of the smallest angle in the triangle.

7. Two forces of 25 newtons and 85 newtons acting on a body form an angle of 55°. Find the magnitude of the resultant force, to the *nearest hundredth of a newton*. Find the measure, to the *nearest degree*, of the angle formed between the resultant and the larger force.

8. In $\triangle ABC$, m$\angle A$ = 32, a = 12, and b = 10. Find the measures of the missing angles and side of $\triangle ABC$. Round each measure to the *nearest tenth*.

VI. Probability

1. A dartboard is shown in the diagram below. The two lines intersect at the center of the circle, and the central angle in sector 2 measures $\frac{2\pi}{3}$.

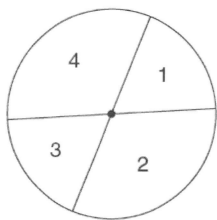

If darts thrown at this board are equally likely to land anywhere on the board, what is the probability that a dart that hits the board will land in either sector 1 or sector 3?
(1) $\frac{1}{6}$ (2) $\frac{1}{3}$ (3) $\frac{1}{2}$ (4) $\frac{2}{3}$

2. A four-digit serial number is to be created from the digits 0 through 9.
How many of these serial numbers can be created if 0 can *not* be the first digit, no digit may be repeated, and the last digit must be 5?
(1) 448 (2) 504 (3) 2,240 (4) 2,520

3. Which formula can be used to determine the total number of different eight-letter arrangements that can be formed using the letters in the word *DEADLINE*?
(1) $8!$ (2) $\frac{8!}{4!}$ (3) $\frac{8!}{2!+2!}$ (4) $\frac{8!}{2! \cdot 2!}$

4. Twenty different cameras will be assigned to several boxes. Three cameras will be randomly selected and assigned to box A. Which expression can be used to calculate the number of ways that three cameras can be assigned to box A?
(1) $20!$ (2) $\frac{20!}{3!}$ (3) $_{20}C_3$ (4) $_{20}P_3$

5. The principal would like to assemble a committee of 8 students from the 15-member student council. How many different committees can be chosen?
(1) 120 (2) 6,435 (3) 32,432,400 (4) 259,459,200

Practice   **VI. Probability**

6. Three marbles are to be drawn at random, without replacement, from a bag containing 15 red marbles, 10 blue marbles, and 5 white marbles. Which expression can be used to calculate the probability of drawing 2 red marbles and 1 white marble from the bag?

(1) $\dfrac{{}_{15}C_2 \cdot {}_5C_1}{{}_{30}C_3}$ (2) $\dfrac{{}_{15}P_2 \cdot {}_5P_1}{{}_{30}C_3}$ (3) $\dfrac{{}_{15}C_2 \cdot {}_5C_1}{{}_{30}P_3}$ (4) $\dfrac{{}_{15}P_2 \cdot {}_5P_1}{{}_{30}P_3}$

7. What is the fourth term in the expansion of $(3x-2)^5$?
(1) $-720x^2$ (2) $-240x$ (3) $720x^2$ (4) $1,080x^3$

Show Work:

1. The letters of any word can be rearranged. Carol believes that the number of different 9-letter arrangements of the word "TENNESSEE" is greater than the number of different 7-letter arrangements of the word "VERMONT." Is she correct? Justify your answer.

2. Find the total number of different twelve-letter arrangements that can be formed using the letters in the word *PENNSYLVANIA*.

3. A committee of 5 members is to be randomly selected from a group of 9 teachers and 20 students. Determine how many different committees can be formed if 2 members must be teachers and 3 members must be students.

VI. Probability

4. The members of a men's club have a choice of wearing black or red vests to their club meetings. A study done over a period of many years determined that the percentage of black vests worn is 60%. If there are 10 men at a club meeting on a given night, what is the probability, to the *nearest thousandth*, that *at least* 8 of the vests worn will be black?

5. The probability that the Stormville Sluggers will win a baseball game is $\frac{2}{3}$. Determine the probability, to the *nearest thousandth*, that the Stormville Sluggers will win *at least* 6 of their next 8 games.

6. A study shows that 35% of the fish caught in a local lake had high levels of mercury. Suppose that 10 fish were caught from this lake. Find, to the *nearest tenth of a percent*, the probability that *at least* 8 of the 10 fish caught did *not* contain high levels of mercury.

7. Write the binomial expansion of $(2x - 1)^5$ as a polynomial in simplest form.

Practice   VII. Statistics

1. A survey completed at a large university asked 2,000 students to estimate the average number of hours they spend studying each week. Every tenth student entering the library was surveyed. The data showed that the mean number of hours that students spend studying was 15.7 per week. Which characteristic of the survey could create a bias in the results?
(1) the size of the sample
(2) the size of the population
(3) the method of analyzing the data
(4) the method of choosing the students who were surveyed

2. Which task is *not* a component of an observational study?
(1) The researcher decides who will make up the sample.
(2) The researcher analyzes the data received from the sample.
(3) The researcher gathers data from the sample, using surveys or taking measurements.
(4) The researcher divides the sample into two groups, with one group acting as a control group.

3. The table below shows the first-quarter averages for Mr. Harper's statistics class.

Statistics Class Averages

Quarter Averages	Frequency
99	1
97	5
95	4
92	4
90	7
87	2
84	6
81	2
75	1
70	2
65	1

What is the population variance for this set of data?
(1)   8.2   (2)   8.3   (3)   67.3   (4)   69.3

4. The lengths of 100 pipes have a normal distribution with a mean of 102.4 inches and a standard deviation of 0.2 inch. If one of the pipes measures exactly 102.1 inches, its length lies
(1) below the 16th percentile
(2) between the 50th and 84th percentiles
(3) between the 16th and 50th percentiles
(4) above the 84th percentile

5. An amateur bowler calculated his bowling average for the season. If the data are normally distributed, about how many of his 50 games were within one standard deviation of the mean?
(1)   14   (2)   17   (3)   34   (4)   48

VII. Statistics

6. Which value of r represents data with a strong negative linear correlation between two variables?
(1) -1.07 (2) -0.89 (3) -0.14 (4) 0.92

Show Work:

1. Howard collected fish eggs from a pond behind his house so he could determine whether sunlight had an effect on how many of the eggs hatched. After he collected the eggs, he divided them into two tanks. He put both tanks outside near the pond, and he covered one of the tanks with a box to block out all sunlight. State whether Howard's investigation was an example of a controlled experiment, an observation, or a survey. Justify your response.

2. Assume that the ages of first-year college students are normally distributed with a mean of 19 years and standard deviation of 1 year.

To the *nearest integer*, find the percentage of first-year college students who are between the ages of 18 years and 20 years, inclusive.

To the *nearest integer*, find the percentage of first-year college students who are 20 years old or older.

3. The scores of one class on the Unit 2 mathematics test are shown in the table below.

Unit 2 Mathematics Test

Test Score	Frequency
96	1
92	2
84	5
80	3
76	6
72	3
68	2

Find the population standard deviation of these scores, to the *nearest tenth*.

Practice — VII. Statistics

4. The table below shows the results of an experiment involving the growth of bacteria.

Time (x) (in minutes)	1	3	5	7	9	11
Number of Bacteria (y)	2	25	81	175	310	497

Write a power regression equation for this set of data, rounding all values to *three decimal places*. Using this equation, predict the bacteria's growth, to the *nearest integer*, after 15 minutes.

5. The table below shows the number of new stores in a coffee shop chain that opened during the years 1986 through 1994.

Year	Number of New Stores
1986	14
1987	27
1988	48
1989	80
1990	110
1991	153
1992	261
1993	403
1994	681

Using $x = 1$ to represent the year 1986 and y to represent the number of new stores, write the exponential regression equation for these data. Round all values to the *nearest thousandth*.

I. Algebraic Expressions, Equations, and Inequalities

1. Factoring Polynomials

1. Factored completely, the expression $6x - x^3 - x^2$ is equivalent to
(1) $x(x+3)(x-2)$
(2) $x(x-3)(x+2)$
(3) $-x(x-3)(x+2)$
*(4) $-x(x+3)(x-2)$

> $6x - x^3 - x^2 = -x^3 - x^2 + 6x = -x(x^2 + x - 6) = -x(x+3)(x-2)$

2. Factored completely, the expression $12x^4 + 10x^3 - 12x^2$ is equivalent to
(1) $x^2(4x+6)(3x-2)$
(2) $2(2x^2+3x)(3x^2-2x)$
(3) $2x^2(2x-3)(3x+2)$
*(4) $2x^2(2x+3)(3x-2)$

> Factor out the common factors first:
> $12x^4 + 10x^3 - 12x^2 = 2x^2(6x^2 + 5x - 6) = 2x^2(2x+3)(3x-2)$

2. Quadratic Equations and Inequalities

3. If $x^2 + 2 = 6x$ is solved by completing the square, an intermediate step would be
(1) $(x+3)^2 = 7$
*(2) $(x-3)^2 = 7$
(3) $(x-3)^2 = 11$
(4) $(x-6)^2 = 34$

> $x^2 - 6x + 2 = 0$
> $x^2 - 6x + (\frac{-6}{2})^2 = -2 + (\frac{-6}{2})^2$
> $x^2 - 6x + (-3)^2 = -2 + (-3)^2$
> $(x-3)^2 = 7$

4. The solutions of the equation $y^2 - 3y = 9$ are
(1) $\frac{3 \pm 3i\sqrt{3}}{2}$
(2) $\frac{3 \pm 3i\sqrt{5}}{2}$
(3) $\frac{-3 \pm 3\sqrt{5}}{2}$
*(4) $\frac{3 \pm 3\sqrt{5}}{2}$

> $y^2 - 3y - 9 = 0$, use Quadratic Formula: $a = 1, b = -3, c = -9$
> $y = \frac{-b \pm \sqrt{b^2 - 4ac}}{2a} = \frac{3 \pm \sqrt{9 + 36}}{2} = \frac{3 \pm 3\sqrt{5}}{2}$

Answers   I. Algebraic Expressions, Equations, and Inequalities

5. The roots of the equation $2x^2 + 7x - 3 = 0$ are

(1) $-\frac{1}{2}$ and -3 *(3) $\frac{-7 \pm \sqrt{73}}{4}$

(2) $\frac{1}{2}$ and 3 (4) $\frac{7 \pm \sqrt{73}}{4}$

$2x^2 + 7x - 3 = 0$, use Quadratic Formula: a = 2, b = 7, c = -3

$$x = \frac{-b \pm \sqrt{b^2 - 4ac}}{2a} = \frac{-7 \pm \sqrt{49 - 4(2)(-3)}}{2 \cdot 2} = \frac{-7 \pm \sqrt{73}}{4}$$

---

6. The solution set of the inequality $x^2 - 3x > 10$ is

(1) $\{x \mid -2 < x < 5\}$ *(3) $\{x \mid x < -2 \text{ or } x > 5\}$
(2) $\{x \mid 0 < x < 3\}$ (4) $\{x \mid x < -5 \text{ or } x > 2\}$

$x^2 - 3x - 10 > 0$
Solve $x^2 - 3x - 10 = 0$
$(x + 2)(x - 5) = 0$
$x_1 = -2$, $x_2 = 5$, divide the number line by these values.

    (true) $x = 0$ (false)   (true)
⟵———|———————|———⟶
    $x_1 = -2$ $x_2 = 5$

test $x = 0$, $0^2 - 3(0) - 10 > 0$ is false. Therefore its next intervals are true.
$x < -2$ or $x > 5$

---

7. Which values of x are in the solution set of the following system of equations?

$$y = 3x - 6$$
$$y = x^2 - x - 6$$

(1) $0, -4$ *(2) $0, 4$ (3) $6, -2$ (4) $-6, 2$

$x^2 - x - 6 = 3x - 6$
$\quad x^2 - 4x = 0$
$\quad x(x - 4) = 0$
$\quad x = 0, \ x = 4$

3. Rational Expressions and Operations

8. The expression $\dfrac{a^2 b^{-3}}{a^{-4} b^2}$ is equivalent to

*(1) $\dfrac{a^6}{b^5}$ (2) $\dfrac{b^5}{a^6}$ (3) $\dfrac{a^2}{b}$ (4) $a^{-2} b^{-1}$

$$\frac{a^2 b^{-3}}{a^{-4} b^2} = \frac{a^2 \cdot a^4}{b^3 \cdot b^2} = \frac{a^6}{b^5}$$

9. Written in simplest form, the expression $\dfrac{\dfrac{x}{4} - \dfrac{1}{x}}{\dfrac{1}{2x} + \dfrac{1}{4}}$ is equivalent to

(1) $x - 1$ *(2) $x - 2$ (3) $\dfrac{x-2}{2}$ (4) $\dfrac{x^2 - 4}{x + 2}$

$$\frac{\dfrac{x}{4} - \dfrac{1}{x}}{\dfrac{1}{2x} + \dfrac{1}{4}} = \frac{\dfrac{x^2 - 4}{4x}}{\dfrac{2 + x}{4x}} = \frac{x^2 - 4}{2 + x} = \frac{(x+2)(x-2)}{(x+2)} = x - 2$$

10. When $x^{-1} - 1$ is divided by $x - 1$, the quotient is

(1) -1 *(2) $-\dfrac{1}{x}$ (3) $\dfrac{1}{x^2}$ (4) $\dfrac{1}{(x-1)^2}$

$$\frac{\dfrac{1}{x} - 1}{x - 1} = \frac{\dfrac{1-x}{x}}{x - 1} = \frac{(1-x)}{x} \cdot \frac{1}{(x-1)} = -\frac{1}{x} \qquad \text{Hint: } (1 - x) = -(x - 1)$$

11. When $\dfrac{3}{2}x^2 - \dfrac{1}{4}x - 4$ is subtracted from $\dfrac{5}{2}x^2 - \dfrac{3}{4}x + 1$, the difference is

(1) $-x^2 + \dfrac{1}{2}x - 5$ *(2) $x^2 - \dfrac{1}{2}x + 5$ (3) $-x^2 - x - 3$ (4) $x^2 - x - 3$

$$\frac{5}{2}x^2 - \frac{3}{4}x + 1 - (\frac{3}{2}x^2 - \frac{1}{4}x - 4)$$
$$= \frac{5}{2}x^2 - \frac{3}{4}x + 1 - \frac{3}{2}x^2 + \frac{1}{4}x + 4 \qquad \text{combine like terms}$$
$$= x^2 - \frac{1}{2}x + 5$$

Answers   I. Algebraic Expressions, Equations, and Inequalities

4. Radicals

12. The expression $4ab\sqrt{2b} - 3a\sqrt{18b^3} + 7ab\sqrt{6b}$ is equivalent to
    (1) $2ab\sqrt{6b}$
    (2) $16ab\sqrt{2b}$
    (3) $-5ab + 7ab\sqrt{6b}$
    *(4) $-5ab\sqrt{2b} + 7ab\sqrt{6b}$

$4ab\sqrt{2b} - 3a\sqrt{18b^3} + 7ab\sqrt{6b} = 4ab\sqrt{2b} - 9ab\sqrt{2b} + 7ab\sqrt{6b} = -5ab\sqrt{2b} + 7ab\sqrt{6b}$

13. The product of $(3 + \sqrt{5})$ and $(3 - \sqrt{5})$ is
    (1) $4 - 6\sqrt{5}$
    (2) $14 - 6\sqrt{5}$
    (3) 14
    *(4) 4

$(3 + \sqrt{5})(3 - \sqrt{5}) = 3^2 - (\sqrt{5})^2 = 9 - 5 = 4$ Hint: $(a + b)(a - b) = a^2 - b^2$

14. If $a = 3$ and $b = -2$, what is the value of the expression $\dfrac{a^{-2}}{b^{-3}}$?
    (1) $-\dfrac{9}{8}$
    (2) -1
    *(3) $-\dfrac{8}{9}$
    (4) $\dfrac{8}{9}$

$\dfrac{a^{-2}}{b^{-3}} = \dfrac{b^3}{a^2} = \dfrac{(-2)^3}{3^2} = \dfrac{-8}{9} = -\dfrac{8}{9}$

15. The expression $(x^2 - 1)^{-\frac{2}{3}}$ is equivalent to
    (1) $\sqrt[3]{(x^2 - 1)^2}$
    *(2) $\dfrac{1}{\sqrt[3]{(x^2 - 1)^2}}$
    (3) $\sqrt{(x^2 - 1)^3}$
    (4) $\dfrac{1}{\sqrt{(x^2 - 1)^3}}$

$(x^2 - 1)^{-\frac{2}{3}} = \dfrac{1}{(x^2 - 1)^{\frac{2}{3}}} = \dfrac{1}{\sqrt[3]{(x^2 - 1)^2}}$

16. When simplified, the expression $\left(\dfrac{w^{-5}}{w^{-9}}\right)^{\frac{1}{2}}$ is equivalent to
    (1) w^{-7}
    *(2) w^2
    (3) w^7
    (4) w^{14}

$\left(\dfrac{w^{-5}}{w^{-9}}\right)^{\frac{1}{2}} = \left(\dfrac{w^9}{w^5}\right)^{\frac{1}{2}} = (x^4)^{\frac{1}{2}} = w^2$

I. Algebraic Expressions, Equations, and Inequalities

17. The expression $x^{-\frac{2}{5}}$ is equivalent to

    (1) $-\sqrt[2]{x^5}$ (2) $-\sqrt[5]{x^2}$ (3) $\dfrac{1}{\sqrt[2]{x^5}}$ *(4) $\dfrac{1}{\sqrt[5]{x^2}}$

$$x^{-\frac{2}{5}} = \frac{1}{x^{\frac{2}{5}}} = \frac{1}{\sqrt[5]{x^2}}$$

18. Which expression is equivalent to $\dfrac{\sqrt{3}+5}{\sqrt{3}-5}$?

    *(1) $-\dfrac{14+5\sqrt{3}}{11}$ (2) $-\dfrac{17+5\sqrt{3}}{11}$ (3) $\dfrac{14+5\sqrt{3}}{14}$ (4) $\dfrac{17+5\sqrt{3}}{14}$

$$\frac{(\sqrt{3}+5)}{(\sqrt{3}-5)} \cdot \frac{(\sqrt{3}+5)}{(\sqrt{3}+5)} = \frac{\sqrt{3}\cdot\sqrt{3}+5\sqrt{3}+5\sqrt{3}+5\cdot 5}{(\sqrt{3})^2-5^2} = \frac{28+10\sqrt{5}}{-22} = -\frac{14+5\sqrt{3}}{11}$$

19. The fraction $\dfrac{3}{\sqrt{3a^2 b}}$ is equivalent to

    (1) $\dfrac{1}{a\sqrt{b}}$ (2) $\dfrac{\sqrt{b}}{ab}$ *(3) $\dfrac{\sqrt{3b}}{ab}$ (4) $\dfrac{\sqrt{3}}{a}$

$$\frac{3}{\sqrt{3a^2b}} = \frac{3}{a\sqrt{3b}} = \frac{3}{a\sqrt{3b}} \cdot \frac{\sqrt{3b}}{\sqrt{3b}} = \frac{3\sqrt{3b}}{3ab} = \frac{\sqrt{3b}}{ab}$$

20. The expression $\dfrac{2x+4}{\sqrt{x+2}}$ is equivalent to

    (1) $\dfrac{(2x+4)\sqrt{x-2}}{x-2}$ (2) $\dfrac{(2x+4)\sqrt{x-2}}{x-4}$ (3) $2\sqrt{x-2}$ *(4) $2\sqrt{x+2}$

$$\frac{2x+4}{\sqrt{x+2}} = \frac{2x+4}{\sqrt{x+2}} \cdot \frac{\sqrt{x+2}}{\sqrt{x+2}} = \frac{2(x+2)\sqrt{x+2}}{(x+2)} = 2\sqrt{x+2}$$

Answers I. Algebraic Expressions, Equations, and Inequalities

21. The solution set of the equation $\sqrt{x+3} = 3-x$ is
*(1) {1} (2) {0} (3) {1, 6} (4) {2, 3}

> Plug in and check.

5. Absolute Value Equations and Inequalities

22. Which graph represents the solution set of $|6x - 7| \leq 5$?

*(1) [number line with solid segment from $\frac{1}{3}$ to 2]
(2) [number line]
(3) [number line]
(4) [number line]

> $-5 \leq 6x - 7 \leq 5$
>
> $-5 \leq 6x - 7$ $6x - 7 \leq 5$
> $2 \leq 6x$ $6x \leq 12$
> $\frac{1}{3} \leq x$ $x \leq 2$
>
> $\frac{1}{3} \leq x \leq 2$

23. What is the solution set of the equation $|4a + 6| - 4a = -10$?
*(1) ∅ (2) {0} (3) $\{\frac{1}{2}\}$ (4) $\{0, \frac{1}{2}\}$

> Plug in the numbers and check.
> **No solutions.**

6. Complex Numbers

24. In simplest form, $\sqrt{-300}$ is equivalent to
(1) $3i\sqrt{10}$ (2) $5i\sqrt{12}$ *(3) $10i\sqrt{3}$ (4) $12i\sqrt{5}$

> $\sqrt{-300} = i\sqrt{300} = 10i\sqrt{3}$

25. The product of i^7 and i^5 is equivalent to
*(1) 1 (2) -1 (3) i (4) $-i$

> $i^7 \cdot i^5 = i^{12} = i^0 = 1$

I. Algebraic Expressions, Equations, and Inequalities

26. The expression $2i^2 + 3i^3$ is equivalent to
*(1) $-2 - 3i$ (2) $2 - 3i$ (3) $-2 + 3i$ (4) $2 + 3i$

> $i^2 = -1$, $i^3 = -i$

27. The expression $(3 - 7i)^2$ is equivalent to
(1) $-40 + 0i$ *(2) $-40 - 42i$ (3) $58 + 0i$ (4) $58 - 42i$

> (3 - 7i)(3 - 7i) = 3•3 + 3•(-7i) + (-7i)•3 + (-7i)(-7i) = 9 - 21i - 21i - 49 = -40 - 42i

28. What is the conjugate of $-2 + 3i$?
(1) $-3 + 2i$ *(2) $-2 - 3i$ (3) $2 - 3i$ (4) $3 + 2i$

> a + bi and a - bi are conjugates

29. The conjugate of $7 - 5i$ is
(1) $-7 - 5i$ (2) $-7 + 5i$ (3) $7 - 5i$ *(4) $7 + 5i$

> a + bi and a - bi are conjugates.

7. Roots of the Quadratic Equations

30. For which equation does the sum of the roots equal $\frac{3}{4}$ and the product of the roots equal -2?
(1) $4x^2 - 8x + 3 = 0$ *(3) $4x^2 - 3x - 8 = 0$
(2) $4x^2 + 8x + 3 = 0$ (4) $4x^2 + 3x - 2 = 0$

> Eq.(3) $4x^2 - 3x - 8 = 0$ is $x^2 - \frac{3}{4}x - 2 = 0$
>
> Here $x_1 + x_2 = -\frac{b}{a} = \frac{3}{4}$, $x_1 \cdot x_2 = \frac{c}{a} = -2$

31. The roots of the equation $9x^2 + 3x - 4 = 0$ are
(1)   imaginary   (3)   real, rational, and unequal
(2)   real, rational, and equal   *(4)   real, irrational, and unequal

> $b^2 - 4ac = 3^2 - 4 \cdot 9 \cdot (-4) = 9 + 144 = 153$
> 153 is greater than 0. The equation has two unequal real roots.
> 153 is not a perfect square. $\sqrt{153}$ is irrational.

Answers **I. Algebraic Expressions, Equations, and Inequalities**

32. The roots of the equation $x^2 - 10x + 25 = 0$ are
(1) imaginary *(3) real, rational, and equal
(2) real and irrational (4) real, rational, and unequal

> Discriminant: $b^2 - 4ac = (-10)^2 - 4 \cdot 1 \cdot 25 = 100 - 100 = 0$

33. For which equation does the sum of the roots equal -3 and the product of the roots equal 2?
(1) $x^2 + 2x - 3 = 0$ *(3) $2x^2 + 6x + 4 = 0$
(2) $x^2 - 3x + 2 = 0$ (4) $2x^2 - 6x + 4 = 0$

> Eq. (3) $2x^2 + 6x + 4 = 0$ is $x^2 + 3x + 2 = 0$
> Here $x_1 + x_2 = -\dfrac{b}{a} = -3$ and $x_1 \cdot x_2 = \dfrac{c}{a} = 2$

8. Polynomial Equations of Higher Degrees

34. The graph of $y = f(x)$ is shown below.

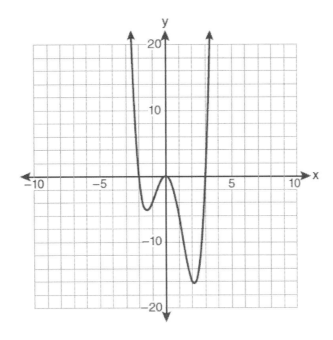

Which set lists all the real solutions of $f(x) = 0$?
(1) $\{-3, 2\}$ (2) $\{-2, 3\}$ (3) $\{-3, 0, 2\}$ *(4) $\{-2, 0, 3\}$

> The real solutions are the values of x in the x-intercepts (-2, 0), (0, 0) and (3, 0).

I. Algebraic Expressions, Equations, and Inequalities

35. The graph of $y = x^3 - 4x^2 + x + 6$ is shown below.

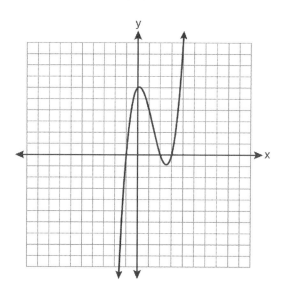

What is the product of the roots of the equation $x^3 - 4x^2 + x + 6 = 0$?
(1) -36 *(2) -6 (3) 6 (4) 4

The roots of an equation are the x values of the x-intercepts (-1, 0), (2, 0) and (3, 0).
Here $x_1 = -1$, $x_2 = 2$, $x_3 = 3$. $x_1 \cdot x_2 \cdot x_3 = (-1) \cdot 2 \cdot 3 = -6$

---

36. Which values of x are solutions of the equation $x^3 + x^2 - 2x = 0$?
(1) 0, 1, 2   (3) 0, −1, 2
*(2) 0, 1, −2   (4) 0, −1, −2

$x(x^2 + x - 2) = 0$
$x(x-1)(x+2) = 0$
$x = 0$, $x = 1$, $x = -2$

---

9. Sequence and Series

37. The value of the expression $2\sum_{n=0}^{2}(n^2 + 2^n)$ is
(1) 12   (2) 22   *(3) 24   (4) 26

$2[(0^2 + 2^0) + (1^2 + 2^1) + (2^2 + 2^2)] = 2(0 + 1 + 1 + 2 + 4 + 4) = 24$

Answers I. Algebraic Expressions, Equations, and Inequalities

38. Mrs. Hill asked her students to express the sum $1 + 3 + 5 + 7 + 9 + \ldots + 39$ using sigma notation. Four different student answers were given. Which student answer is correct?

*(1) $\sum_{k=1}^{20}(2k-1)$ (2) $\sum_{k=2}^{40}(k-1)$ (3) $\sum_{k=-1}^{37}(k+2)$ (4) $\sum_{k=1}^{39}(2k-1)$

> There are 20 terms.

39. What is the common difference of the arithmetic sequence $5, 8, 11, 14$?

(1) $\dfrac{8}{5}$ (2) -3 *(3) 3 (4) 9

40. What is a formula for the nth term of sequence B shown below?
$$B = 10, 12, 14, 16, \ldots$$

*(1) $b_n = 8 + 2n$ (2) $b_n = 10 + 2n$ (3) $b_n = 10(2)^n$ (4) $b_n = 10(2)^{n-1}$

> This is an arithmetic sequence with the common difference 2.
> For $b_n = 8 + 2n$, $b_1 = 8 + 2\cdot 1 = 10$, $b_2 = 8 + 2\cdot 2 = 12$, $b_3 = 8 + 2\cdot 3 = 14$, ...

41. Which arithmetic sequence has a common difference of 4?
(1) $\{0, 4n, 8n, 12n, \ldots\}$
(2) $\{n, 4n, 16n, 64n, \ldots\}$
*(3) $\{n+1, n+5, n+9, n+13, \ldots\}$
(4) $\{n+4, n+16, n+64, n+256, \ldots\}$

42. What is the formula for the nth term of the sequence $54, 18, 6, \ldots$?

(1) $a_n = 6\left(\dfrac{1}{3}\right)^n$ (2) $a_n = 6\left(\dfrac{1}{3}\right)^{n-1}$ (3) $a_n = 54\left(\dfrac{1}{3}\right)^n$ *(4) $a_n = 54\left(\dfrac{1}{3}\right)^{n-1}$

> This is a geometric sequence with the common ratio $\dfrac{1}{3}$.
> For $a_n = 54(\dfrac{1}{3})^{n-1}$, $a_1 = 54(\dfrac{1}{3})^0 = 54$, $a_2 = 54(\dfrac{1}{3})^1 = 18$, $a_3 = 54(\dfrac{1}{3})^2 = 6$, ...

43. What is the common ratio of the geometric sequence whose first term is 27 and fourth term is 64?

(1) $\dfrac{3}{4}$ (2) $\dfrac{64}{81}$ *(3) $\dfrac{4}{3}$ (4) $\dfrac{37}{3}$

> Geometric sequence: $a_n = a_1 \cdot r^{n-1}$
> $a_1 = 27$, $a_4 = a_1 \cdot r^{4-1}$, $64 = 27 \cdot r^3$, $r^3 = \dfrac{64}{27}$, $r = \dfrac{4}{3}$

70.   **I. Algebraic Expressions, Equations, and Inequalities**

44. What is the fifteenth term of the sequence 5, −10, 20, −40, 80, . . . ?
(1) −163,840   (2) −81,920   *(3) 81,920   (4) 327,680

$a_n = 5 \cdot (-2)^{n-1}$
$a_1 = 5 \cdot (-2)^{1-1} = 5$, $a_2 = 5 \cdot (-2)^{2-1} = -10$, $a_3 = 5 \cdot (-2)^{3-1} = 20$, $a_4 = 5 \cdot (-2)^{3-1} = -40$
$a_{15} = 5 \cdot (-2)^{15-1} = \mathbf{81{,}920}$

Show Work:

Polynomials, Rationals, Radicals, Absolute Values

1. Factor completely: $10ax^2 - 23ax - 5a$

$= a(10x^2 - 23x - 5)$
$= a(5x + 1)(2x - 5)$

$5 \cdot 2 = 10$ $\begin{matrix} 5 \\ 2 \end{matrix} \times \begin{matrix} 1 \\ -5 \end{matrix}$ $1 \cdot (-5) = -5$

$5 \cdot (-5) + 2 \cdot 1 = -23$

---

2. Solve $2x^2 - 12x + 4 = 0$ by completing the square, expressing the result in simplest radical form.

$x^2 - 6x + 2 = 0$ simplify in standard form $x^2 + bx + c = 0$
$x^2 - 6x = -2$ move the constant c to the right side of the equation and change its sign
$x^2 - 6x + \left(\dfrac{-6}{2}\right)^2 = -2 + \left(\dfrac{-6}{2}\right)^2$ add $\left(\dfrac{b}{2}\right)^2$ on both sides of the equation
$(x - 3)^2 = 7$ the left side is the perfect square of $\left(x + \dfrac{b}{2}\right)^2$
$x - 3 = \pm\sqrt{7}$
$\mathbf{x = 3 \pm \sqrt{7}}$

---

3. Express in simplest form: $\dfrac{\dfrac{1}{2} - \dfrac{4}{d}}{\dfrac{1}{d} + \dfrac{3}{2d}}$

$= \dfrac{\dfrac{d-8}{2d}}{\dfrac{2+3}{2d}} = \dfrac{d-8}{2d} \cdot \dfrac{2d}{2+3} = \dfrac{d-8}{5}$

Answers   I. Algebraic Expressions, Equations, and Inequalities

4. Solve for x: $\dfrac{4x}{x-3} = 2 + \dfrac{12}{x-3}$

$4x = 2(x - 3) + 12$ multiply LCD: $x - 3$ on both sides of the equation
$4x = 2x - 6 + 12$
$2x = 6$
$x = 3$ check: undefined
No solution.

---

5. Solve algebraically for x: $\dfrac{1}{x+3} - \dfrac{2}{3-x} = \dfrac{4}{x^2-9}$

Rewrite the equation as:
$\dfrac{1}{x+3} + \dfrac{2}{x-3} = \dfrac{4}{x^2-9}$ here $3 - x = -(x - 3)$

$x - 3 + 2(x + 3) = 4$ multiply LCD: $(x + 3)(x - 3)$ on both sides of the equation
$x - 3 + 2x + 6 = 4$
$3x = 1$
$x = \dfrac{1}{3}$ check: OK

---

6. Express $5\sqrt{3x^3} - 2\sqrt{27x^3}$ in simplest radical form.

$= 5x\sqrt{3x} - 2 \bullet 3x\sqrt{3x}$
$= -x\sqrt{3x}$

---

7. Express $\left(\dfrac{2}{3}x - 1\right)^2$ as a trinomial.

$= \left(\dfrac{2}{3}x - 1\right)\left(\dfrac{2}{3}x - 1\right)$
$= \left(\dfrac{2}{3}x\right)^2 - \dfrac{2}{3}x - \dfrac{2}{3}x + (-1)^2$
$= \dfrac{4}{9}x^2 - \dfrac{4}{3}x + 1$

I. Algebraic Expressions, Equations, and Inequalities

8. Express $\dfrac{5}{3-\sqrt{2}}$ with a rational denominator, in simplest radical form.

$$= \dfrac{5}{(3-\sqrt{2})} \cdot \dfrac{(3+\sqrt{2})}{(3+\sqrt{2})} = \dfrac{15+5\sqrt{2}}{3^2-(\sqrt{2})^2} = \dfrac{15+5\sqrt{2}}{9-2} = \dfrac{15+5\sqrt{2}}{7}$$

9. Express $\dfrac{\sqrt{108x^5y^8}}{\sqrt{6xy^5}}$ in simplest radical form.

$$= \sqrt{\dfrac{108x^5y^8}{6xy^5}} = \sqrt{18x^4y^3} = 3x^2y\sqrt{2y}$$

Complex Numbers, Polynomial Equations, Roots, Sequence and Series

10. Use the discriminant to determine all value of k that would result in the equation $x^2 - kx + 4 = 0$ having equal roots.

$b^2 - 4ac = 0$
$(-k)^2 - 4 \cdot 1 \cdot 4 = 0$
$k^2 = 16$
$k = \pm 4$

11. Find the sum and product of the roots of the equation $5x^2 + 11x - 3 = 0$.

$a = 5$, $b = 11$, $c = -3$

The sum of the roots $x_1 + x_2 = -\dfrac{b}{a} = -\dfrac{11}{5}$

The product of the roots $x_1 \cdot x_2 = \dfrac{c}{a} = -\dfrac{3}{5}$

Answers   **I. Algebraic Expressions, Equations, and Inequalities**

12. Solve the equation $8x^3 + 4x^2 - 18x - 9 = 0$ algebraically for all values of x.

> Grouping the four-term polynomial:
> $4x^2(2x + 1) - 9(2x + 1) = 0$
> $(2x + 1)(4x^2 - 9) = 0$
> $(2x + 1)(2x + 3)(2x - 3) = 0$
> $2x + 1 = 0, \quad 2x + 3 = 0, \quad 2x - 3 = 0$
> $x = -\dfrac{1}{2}, \quad x = -\dfrac{3}{2}, \quad x = \dfrac{3}{2}$

13. Find the first four terms of the recursive sequence defined below.

$$a_1 = -3$$
$$a_n = a_{(n-1)} - n$$

> $a_2 = a_{(2-1)} - 2 = a_1 - 2 = -3 - 2 = -5$
> $a_3 = a_{(3-1)} - 3 = a_2 - 3 = -5 - 3 = -8$
> $a_4 = a_{(4-1)} - 4 = a_3 - 4 = -8 - 4 = -12$
> **The first four terms: {-3, -5, -8, -12}**

14. Express the sum $7 + 14 + 21 + 28 + \ldots + 105$ using sigma notation.

> The nth term: $a_n = 7n$, $a_1 = 7 \bullet 1 = 7$, $a_{15} = 7 \bullet 15 = 105$
>
> The sum: $\displaystyle\sum_{n=1}^{15} 7n$

15. Evaluate: $10 + \displaystyle\sum_{n=1}^{5} (n^3 - 1)$

> $= 10 + (1^3 - 1) + (2^3 - 1) + (3^3 - 1) + (4^3 - 1) + (5^3 - 1)$
> $= 10 + 0 + 7 + 26 + 63 + 124$
> $= \mathbf{230}$

II. Relations and Functions

1. Which function is *not* one-to-one?
(1) {(0, 1), (1, 2), (2, 3), (3, 4)}
(2) {(0, 0), (1, 1), (2, 2), (3, 3)}
(3) {(0, 1), (1, 0), (2, 3), (3, 2)}
*(4) {(0, 1), (1, 0), (2, 0), (3, 2)}

> One-to-one function has no repeated first element and no repeated second element.
> In choice (4), (1, 0) and (2, 0) have the same second element 0.

2. What is the domain of the function $f(x) = \sqrt{x-2} + 3$?
(1) $(-\infty, \infty)$
(2) $(2, \infty)$
*(3) $[2, \infty)$
(4) $[3, \infty)$

> $x - 2 \geq 0$ the radicand of a square root can not be negative
> $x \geq 2$ expressed as interval notation: $[2, \infty)$

3. What are the domain and the range of the function shown in the graph below?

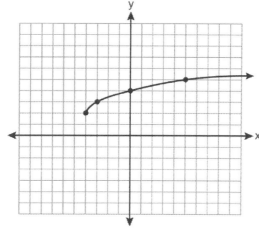

(1) $\{x | x > -4\}$; $\{y | y > 2\}$
*(2) $\{x | x \geq -4\}$; $\{y | y \geq 2\}$
(3) $\{x | x > 2\}$; $\{y | y > -4\}$
(4) $\{x | x \geq 2\}$; $\{y | y \geq -4\}$

4. Which graph does *not* represent a function?
(1)   (2)   (3)   *(4)

> (4) has failed the vertical line test (intersects more than one point).

Answers II. Relations and Functions

5. The equation $y - 2\sin\theta = 3$ may be rewritten as
(1) $f(y) = 2\sin x + 3$
(2) $f(y) = 2\sin\theta + 3$
(3) $f(x) = 2\sin\theta + 3$
*(4) $f(\theta) = 2\sin\theta + 3$

> e.g. $y = x^2$ can be written as $f(x) = x^2$. (y is a function of x)
> Rewrite the equation as $y = 2\sin\theta + 3$.
> It can be written as $f(\theta) = 2\sin\theta + 3$. (y is a function of θ)

6. Which relation is *not* a function?
*(1) $(x-2)^2 + y^2 = 4$
(2) $x^2 + 4x + y = 4$
(3) $x + y = 4$
(4) $xy = 4$

> (1) is an equation of a circle, which is not a function.
> Tips: when an equation contains a term y^2, it is not a function.

7. Which graph represents a one-to-one function?

(1)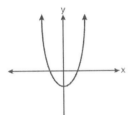

*(3)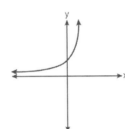

(2)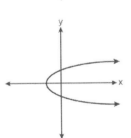

(4)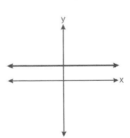

> (3) passes both vertical line test and horizontal line test (intersects the graph at only one point).

76.   II. Relations and Functions

8. Which graph does *not* represent a function?

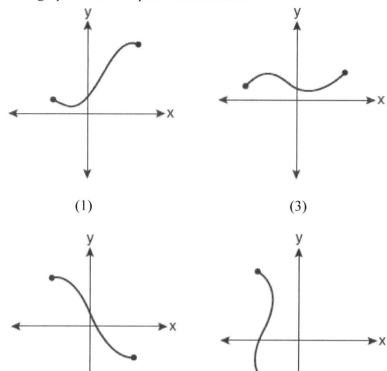

(1) (3)

(2) *(4)

(4) does not pass the vertical line test.

9. If $f(x) = \frac{1}{2}x - 3$ and $g(x) = 2x + 5$, what is the value of $(g \circ f)(4)$?
(1)   −13   (2)   3.5   *(3)   3   (4)   6

$(g \circ f)(4) = g(f(4)) = g(-1) = 3$
Hint: $f(4) = \frac{1}{2} \cdot 4 - 3 = -1$, $g(-1) = 2(-1) + 5 = 3$

10. If $f(x) = x^2 - 5$ and $g(x) = 6x$, then $g(f(x))$ is equal to
(1) $6x^3 - 30x$ (3) $36x^2 - 5$
*(2) $6x^2 - 30$ (4) $x^2 + 6x - 5$

$g(f(x)) = g(x^2 - 5) = 6(x^2 - 5) = 6x^2 - 30$

Answers II. Relations and Functions 77.

11. The minimum point on the graph of the equation $y = f(x)$ is $(-1,-3)$. What is the minimum point on the graph of the equation $y = f(x) + 5$?
*(1) $(-1, 2)$ (2) $(-1,-8)$ (3) $(4,-3)$ (4) $(-6,-3)$

> $y = f(x)$ moved 5 units up $y = f(x) + 5$
>
> The y value of the original minimum point is -3.
> The y value of the new minimum point is -3 + 5 = **2**.
> The x value is not changed. x = **-1**

12. The graph below shows the function $f(x)$.

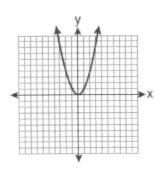

Which graph represents the function $f(x+2)$?
(1)   (3)

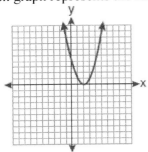   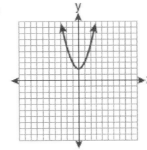

*(2)   (4)

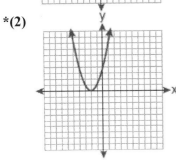   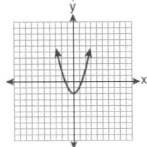

> $f(x)$ moved 2 units to the left $y = f(x + 2)$

78.   II. Relations and Functions

Important Functions and Relations

13. Which graph best represents the inequality $y + 6 \geq x^2 - x$?

*(1)   (3)

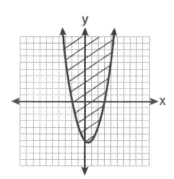
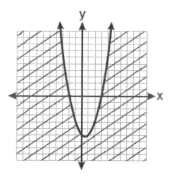

(2)   (4)

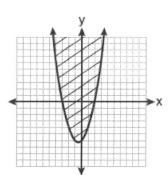
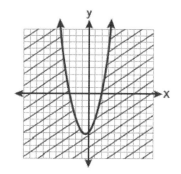

Rewrite the inequality in standard form: $y \geq x^2 - x - 6$

The axis of symmetry: $x = \dfrac{-b}{2a} = \dfrac{1}{2 \cdot 1} = \dfrac{1}{2}$

The vertex: $x = \dfrac{1}{2}$, $y = (\dfrac{1}{2})^2 - (\dfrac{1}{2}) - 6 = \dfrac{1}{4} - \dfrac{1}{2} - 6 = -6\dfrac{1}{4}$, $(\dfrac{1}{2}, -6\dfrac{1}{4})$

The shaded region above the curve and the solid line is the solution set.

---

14. The equation $x^2 + y^2 - 2x + 6y + 3 = 0$ is equivalent to

(1) $(x-1)^2 + (y+3)^2 = -3$ (3) $(x+1)^2 + (y+3)^2 = 7$

*(2) $(x-1)^2 + (y+3)^2 = 7$ (4) $(x+1)^2 + (y+3)^2 = 10$

Completing the square:
$x^2 + y^2 - 2x + 6y + 3 = 0$
$x^2 - 2x + y^2 + 6y = -3$
$x^2 - 2x + (\dfrac{-2}{2})^2 + y^2 + 6y + (\dfrac{6}{2})^2 = -3 + (\dfrac{-2}{2})^2 + (\dfrac{6}{2})^2$
$(x - 1)^2 + (y + 3)^2 = 7$

Answers   II. Relations and Functions

15. Which equation represents the circle shown in the graph below that passes through the point $(0,-1)$?

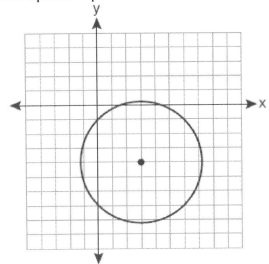

(1) $(x-3)^2 + (y+4)^2 = 16$
*(2) $(x-3)^2 + (y+4)^2 = 18$
(3) $(x+3)^2 + (y-4)^2 = 16$
(4) $(x+3)^2 + (y-4)^2 = 18$

> Center (h, k) is $(3, -4)$
> Radius r is the distance between $(3, -4)$ and $(0, -1)$.
> $r = \sqrt{(x_2 - x_1)^2 + (y_2 - y_1)^2} = \sqrt{(3-0)^2 + (-4+1)^2} = \sqrt{9+9} = \sqrt{18}$
> Equation of Circle: $(x - h)^2 + (y - k)^2 = r^2$

16. Four points on the graph of the function $f(x)$ are shown below.
$$\{(0, 1), (1, 2), (2, 4), (3, 8)\}$$
Which equation represents $f(x)$?

*(1) $f(x) = 2^x$ (2) $f(x) = 2x$ (3) $f(x) = x + 1$ (4) $f(x) = \log_2 x$

> $2^0 = 1,\ 2^1 = 2,\ 2^2 = 4,\ 2^3 = 8$

17. The solution set of $4^{x^2 + 4x} = 2^{-6}$ is
(1) $\{1, 3\}$ (2) $\{-1, 3\}$ *(3) $\{-1, -3\}$ (4) $\{1, -3\}$

> $2^{2(x^2 + 4x)} = 2^{-6}$ change to the same base
> $2(x^2 + 4x) = -6$ solve the equation of the exponents
> $x^2 + 4x = -3$ simplify
> $x^2 + 4x + 3 = 0$ standard form
> $(x + 1)(x + 3) = 0$
> $x = -1,\ x = -3$

II. Relations and Functions

18. What is the value of x in the equation $9^{3x+1} = 27^{x+2}$?

(1) 1   (2) $\frac{1}{3}$ (3) $\frac{1}{2}$ *(4) $\frac{4}{3}$

$3^{2(3x+1)} = 3^{3(x+2)}$ change to the same base
$2(3x+1) = 3(x+2)$ solve the equation of the exponents
$6x + 2 = 3x + 6$
$3x = 4$
$x = \frac{4}{3}$

---

19. On January 1, a share of a certain stock cost $180. Each month thereafter, the cost of a share of this stock decreased by one-third. If x represents the time, in months, and y represents the cost of the stock, in dollars, which graph best represents the cost of a share over the following 5 months?

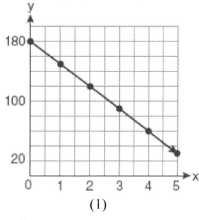
(1)

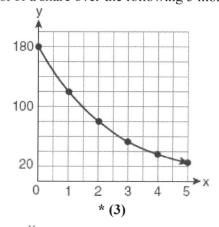
* (3)

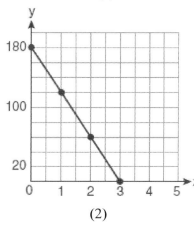
(2)

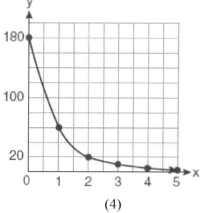
(4)

$y = 180 \cdot (1 - \frac{1}{3})^x$

(1, 120), (2, 80), (3, 53), (4, 36)

Answers   **II. Relations and Functions**

20. The expression $\log_5\left(\dfrac{1}{25}\right)$ is equivalent to

(1) $\dfrac{1}{2}$ (2) 2 (3) $-\dfrac{1}{2}$ *(4) -2

$\log_5\left(\dfrac{1}{5^2}\right) = \log_5(5^{-2}) = \mathbf{-2}$

---

21. The expression $\log_8 64$ is equivalent to

(1) 8 *(2) 2 (3) $\dfrac{1}{2}$ (4) $\dfrac{1}{8}$

Let $\log_8 64 = x$, then $8^x = 64$, $x = \mathbf{2}$

---

22. The expression $2\log x - (3\log y + \log z)$ is equivalent to

*(1) $\log\dfrac{x^2}{y^3 z}$ (2) $\log\dfrac{x^2 z}{y^3}$ (3) $\log\dfrac{2x}{3yz}$ (4) $\log\dfrac{2xz}{3y}$

$2\log x - (3\log y + \log z) = \log x^2 - \log y^3 - \log z = \log\dfrac{x^2}{y^3 z}$

---

23. What is the solution of the equation $2\log_4(5x) = 3$?

(1) 6.4 (2) 2.56 (3) $\dfrac{9}{5}$ *(4) $\dfrac{8}{5}$

$2\log_4(5x) = 3$ can be written as $\log_4(5x)^2 = 3$.
Then rewrite the equation in exponential form:
$4^3 = (5x)^2$, $64 = 25x^2$, $x^2 = \dfrac{64}{25}$, $x = \mathbf{\dfrac{8}{5}}$

($x = -\dfrac{8}{5}$ rejected, the domain of the logarithmic function is the set of positive numbers)

II. Relations and Functions

24. Which two functions are inverse functions of each other?
   (1) $f(x) = \sin x$ and $g(x) = \cos(x)$
   *(3) $f(x) = e^x$ and $g(x) = \ln x$
   (2) $f(x) = 3 + 8x$ and $g(x) = 3 - 8x$
   (4) $f(x) = 2x - 4$ and $g(x) = -\frac{1}{2}x + 4$

> The exponential function $y = e^x$ and the logarithmic function $y = \ln x$ are inverse functions of each other.

25. If a function is defined by the equation $f(x) = 4^x$, which graph represents the inverse of this function?

(1)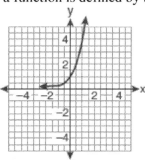

(3)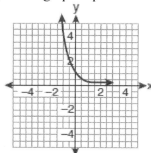

*(2)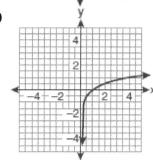

(4)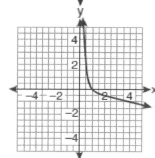

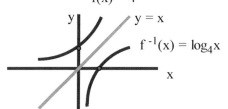

The graph of $f^{-1}(x)$ is the reflection of $f(x)$ in the line $y = x$.

Answers II. Relations and Functions 83.

Show Work:

1. The graph below represents the function $y = f(x)$.

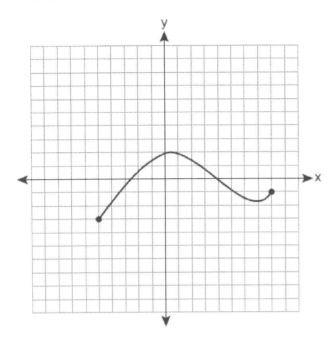

State the domain and range of this function.

> Domain: $\{x \mid -5 \leq x \leq 8\}$ or $[-5, 8]$
> Range: $\{y \mid -3 \leq y \leq 2\}$ or $[-3, 2]$

2. For a given set of rectangles, the length is inversely proportional to the width. In one of these rectangles, the length is 12 and the width is 6. For this set of rectangles, calculate the width of a rectangle whose length is 9.

> Inverse Variation:
> $L_1 \cdot W_1 = L_2 \cdot W_2$
> $12 \cdot 6 = 9 \cdot W_2$
> $W_2 = 8$

3. Write an equation of the circle shown in the graph below.

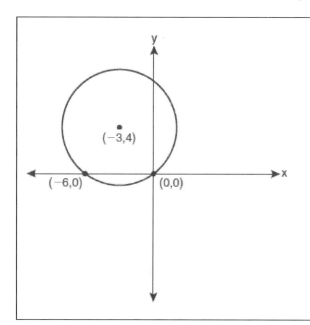

Radius r is the distance between (0, 0) and (-3, 4):
$$r = \sqrt{(x_2 - x_1)^2 + (y_2 - y_1)^2}$$
$$= \sqrt{(-3)^2 + (4)^2}$$
$$= \sqrt{25}$$
$$= 5$$
Center of the circle is (-3, 4).

The center-radius equation of a circle with radius r and center (h, k):
$$(x - h)^2 + (y - k)^2 = r^2$$

$$(x + 3)^2 + (y - 4)^2 = 5^2$$

4. A circle shown in the diagram below has a center of $(-5, 3)$ and passes through point $(-1, 7)$.

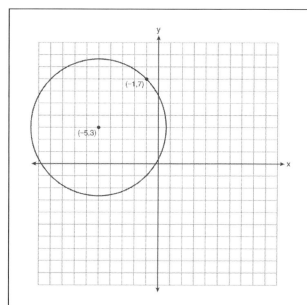

Center: (-5, 3)
Radius $r = \sqrt{(x_2 - x_1)^2 + (y_2 - y_1)^2}$
$$= \sqrt{(-1 + 5)^2 + (7 - 3)^2}$$
$$= \sqrt{16 + 16} = \sqrt{32}$$
$$r^2 = 32$$
$$(x - h)^2 + (y - k)^2 = r^2$$

$$(x + 5)^2 + (y - 3)^2 = 32$$

Write an equation that represents the circle.

Answers   II. Relations and Functions   85.

5. Matt places $1,200 in an investment account earning an annual rate of 6.5%, compounded continuously. Using the formula $V = Pe^{rt}$, where V is the value of the account in t years, P is the principal initially invested, e is the base of a natural logarithm, and r is the rate of interest, determine the amount of money, to the *nearest cent*, that Matt will have in the account after 10 years.

$P = 1,200$, $r = 6.5\% = 0.065$, $t = 10$
$V = 1,200e^{0.065 \cdot 10} = 1,200e^{0.65} =$ **$2,298.65**

6. The graph of the equation $y = \left(\dfrac{1}{2}\right)^x$ has an asymptote. On the grid below, sketch the graph of $y = \left(\dfrac{1}{2}\right)^x$ and write the equation of this asymptote.

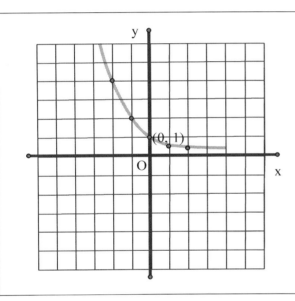

x	-2	-1	0	1	2
y	4	2	1	$\dfrac{1}{2}$	$\dfrac{1}{4}$

x-axis is the horizontal asymptote and its equation is y = 0

7. Solve algebraically for x: $16^{2x+3} = 64^{x+2}$

$4^{2(2x+3)} = 4^{3(x+2)}$
$2(2x+3) = 3(x+2)$
$4x + 6 = 3x + 6$
x = 0

II. Relations and Functions

8. The temperature, T, of a given cup of hot chocolate after it has been cooling for t minutes can best be modeled by the function below, where T_0 is the temperature of the room and k is a constant.
$$\ln(T - T_0) = -kt + 4.718$$

A cup of hot chocolate is placed in a room that has a temperature of 68°. After 3 minutes, the temperature of the hot chocolate is 150°. Compute the value of k to the *nearest thousandth*. [Only an algebraic solution can receive full credit.]

Using this value of k, find the temperature, T, of this cup of hot chocolate if it has been sitting in this room for a total of 10 minutes. Express your answer to the *nearest degree*. [Only an algebraic solution can receive full credit.]

---

(1) $T_0 = 68$, $t = 3$, $T = 150$
$\ln(150 - 68) = -3k + 4.718$
k = 0.104

(2) $T_0 = 68$, $t = 10$, $k = 0.104$
$\ln(T - 68) = -0.104 \cdot 10 + 4.718$
$\ln(T - 68) = 3.678$
$T - 68 = e^{3.678}$
T = 108

---

9. Solve algebraically for x: $\log_{x+3} \dfrac{x^3 + x - 2}{x} = 2$

Rewrite the equation in exponential form:
$(x + 3)^2 = \dfrac{x^3 + x - 2}{x}$

$x^2 + 6x + 9 = \dfrac{x^3 + x - 2}{x}$

$x^3 + 6x^2 + 9x = x^3 + x - 2$
$6x^2 + 8x + 2 = 0$
$3x^2 + 4x + 1 = 0$
$(3x + 1)(x + 1) = 0$
$3x + 1 = 0$, $\quad x + 1 = 0$
$x = -\dfrac{1}{3}$, $\quad x = -1$ (check: both are solutions)

Answers   III. Trigonometric Functions

1. What is the radian measure of an angle whose measure is −420°?

*(1) $-\dfrac{7\pi}{3}$ (2) $-\dfrac{7\pi}{6}$ (3) $\dfrac{7\pi}{6}$ (4) $\dfrac{7\pi}{3}$

$$-420 \cdot \dfrac{\pi}{180} = -\dfrac{7\pi}{3}$$

2. In which graph is θ coterminal with an angle of −70°?

(1)
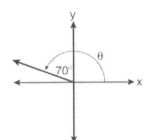

(3)
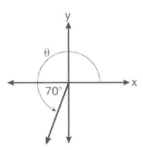

(2)
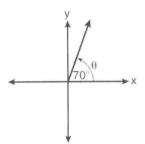

*(4)
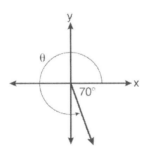

−70° + 360° = 290°, 290° and −70° are coterminal angles.

3. A circle has a radius of 4 inches. In inches, what is the length of the arc intercepted by a central angle of 2 radians?

(1) 2π (2) 2   (3) 8π *(4) 8

$s = r \cdot \theta = 4 \cdot 2 = 8$

4. What is the number of degrees in an angle whose radian measure is $\dfrac{11\pi}{12}$?

(1) 150   *(2) 165   (3) 330   (4) 518

$$\dfrac{11\pi}{12} \cdot \dfrac{180}{\pi} = 165$$

88. III. Trigonometric Functions

5. If ∠A is acute and $\tan A = \frac{2}{3}$, then

(1) $\cot A = \frac{2}{3}$ (2) $\cot A = \frac{1}{3}$ *(3) $\cot(90° - A) = \frac{2}{3}$ (4) $\cot(90° - A) = \frac{1}{3}$

> tanA and cotA are cofunctions:
> tanA = cot(90° - A)

6. In the diagram below of right triangle KTW, KW = 6, KT = 5, and m∠KTW = 90.

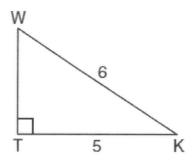

What is the measure of ∠K, to the *nearest minute*?
*(1) 33°33' (2) 33°34' (3) 33°55' (4) 33°56'

> $\cos K = \frac{5}{6}$, $m\angle K = \cos^{-1}(\frac{5}{6}) = 33.5573° = 33°33'$
> convert degree to minute: 0.5573° = 0.5573•60' ≈ 33'

7. Which ratio represents $\csc A$ in the diagram below?

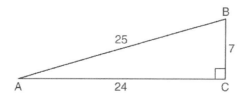

(1) $\frac{25}{24}$ *(2) $\frac{25}{7}$ (3) $\frac{24}{7}$ (4) $\frac{7}{24}$

> $\csc A = \frac{1}{\sin A} = \frac{25}{7}$

8. In the diagram below of right triangle *JTM*, *JT* = 12, *JM* = 6, and m∠*JMT* = 90.

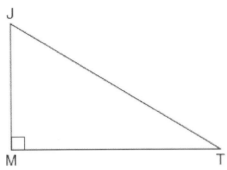

What is the value of cot *J* ?

*(1) $\dfrac{\sqrt{3}}{3}$ (2) 2 (3) $\sqrt{3}$ (4) $\dfrac{2\sqrt{3}}{3}$

$MT = \sqrt{12^2 - 6^2} = \sqrt{108} = 6\sqrt{3}$

$\cot J = \dfrac{6}{6\sqrt{3}} = \dfrac{1}{\sqrt{3}} = \dfrac{1}{\sqrt{3}} \cdot \dfrac{\sqrt{3}}{\sqrt{3}} = \dfrac{\sqrt{3}}{3}$

(or since $\sin T = \dfrac{6}{12} = \dfrac{1}{2}$, m∠T = 30, m∠J = 60, $\cot 60° = \tan 30° = \dfrac{\sqrt{3}}{3}$)

---

9. In the diagram below of a unit circle, the ordered pair $(-\dfrac{\sqrt{2}}{2}, -\dfrac{\sqrt{2}}{2})$ represents the point where the terminal side of θ intersects the unit circle.

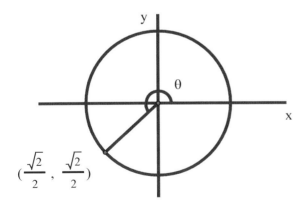

What is m∠θ?
(1)   45   (2)   135   *(3)   225   (4)   240

$\tan\theta = \dfrac{y}{x} = 1$ In Quadrant III m∠θ = 180 + 45 = **225**

90. III. Trigonometric Functions

Show Work:

1. Find, to the *nearest tenth of a degree*, the angle whose measure is 2.5 radians.

$$2.5 \cdot \frac{180°}{\pi} \approx \mathbf{143.2°}$$

2. Find, to the *nearest minute*, the angle whose measure is 3.45 radians.

$$3.45 \cdot \frac{180°}{\pi} \approx 197.6704393° \approx \mathbf{197°40'}$$
$$(0.6704393° = 0.6704393 \cdot 60 \approx 40')$$

3. If θ is an angle in standard position and its terminal side passes through the point $(-3, 2)$, find the exact value of $\csc \theta$.

In general, a point (x, y) is not on a Unit Circle:
$$\sin\theta = \frac{y}{\sqrt{x^2 + y^2}} = \frac{2}{\sqrt{(-3)^2 + 2^2}} = \frac{2}{\sqrt{13}}$$
$$\csc\theta = \frac{1}{\sin\theta} = \frac{\sqrt{13}}{2}$$

4. On the unit circle shown in the diagram below, sketch an angle, in standard position, whose degree measure is 240 and find the exact value of $\sin 240°$.

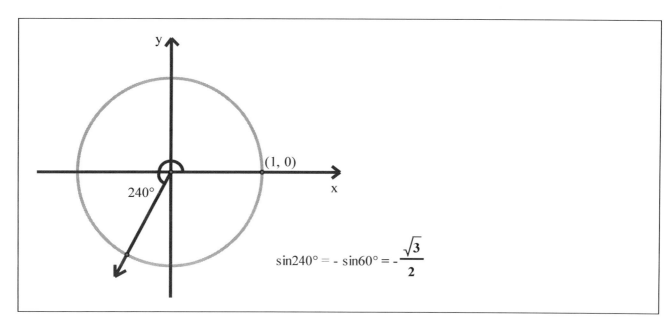

$\sin 240° = -\sin 60° = -\dfrac{\sqrt{3}}{2}$

Answers **IV. Trigonometric Graphs**

1. Which equation is represented by the graph below?

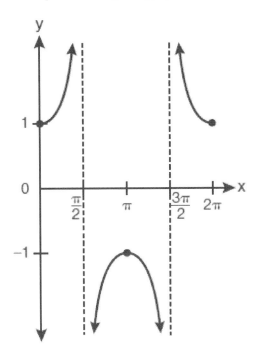

(1) $y = \cot x$ (2) $y = \csc x$ *(3) $y = \sec x$ (4) $y = \tan x$

> $y = \sec x$ is the reciprocal function of $y = \cos x$

2. Which graph represents the equation $y = \cos^{-1} x$?

(1)
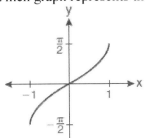

*(3)
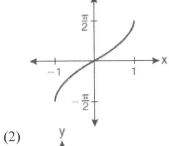

(2)
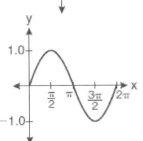

(4)
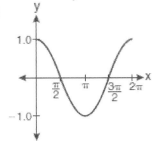

IV. Trigonometric Graphs

3. Which equation is sketched in the diagram below?

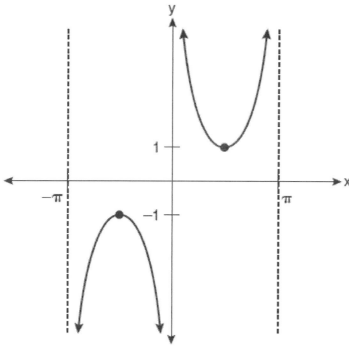

*(1) $y = \csc x$ (2) $y = \sec x$ (3) $y = \cot x$ (4) $y = \tan x$

$y = \csc x = \dfrac{1}{\sin x}$

sin x and csc x are reciprocal functions. Refer to Review page 14.

---

4. What is the period of the function $y = \dfrac{1}{2} \sin\left(\dfrac{x}{3} - \pi\right)$?

(1) $\dfrac{1}{2}$ (2) $\dfrac{1}{3}$ (3) $\dfrac{2}{3}\pi$ *(4) 6π

---

frequency $= \dfrac{1}{3}$,   period $= \dfrac{2\pi}{\text{frequency}} = \dfrac{2\pi}{\frac{1}{3}} = 6\pi$

Answers — IV. Trigonometric Graphs

5. Which graph represents one complete cycle of the equation $y = \sin 3\pi x$?

(1)
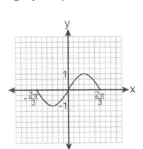

*(3)
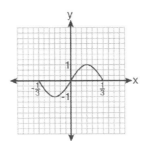

(2)
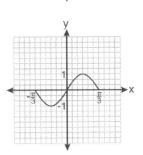

(4)
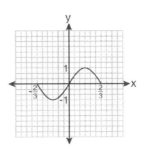

> frequency $b = 3\pi$, period $p = \dfrac{2\pi}{b} = \dfrac{2\pi}{3\pi} = \dfrac{2}{3}$

6. The function $f(x) = \tan x$ is defined in such a way that $f^{-1}(x)$ is a function. What can be the domain of $f(x)$?

(1) $\{x \mid 0 \le x \le \pi\}$ (2) $\{x \mid 0 \le x \le 2\pi\}$ *(3) $\left\{x \mid -\dfrac{\pi}{2} < x < \dfrac{\pi}{2}\right\}$ (4) $\left\{x \mid -\dfrac{\pi}{2} < x < \dfrac{3\pi}{2}\right\}$

> In the restricted domain $\left(-\dfrac{\pi}{2}, \dfrac{\pi}{2}\right)$, $f(x) = \tan x$ is a one-to-one function, therefore its inverse $f^{-1}(x)$ is a function.

7. What is the principal value of $\cos^{-1}\left(-\dfrac{\sqrt{3}}{2}\right)$?

(1) $-30°$ (2) $60°$ *(3) $150°$ (4) $240°$

> The principal value of $\cos^{-1} x$ is in the range $[0°, 180°]$.
> $\cos 150° = -\dfrac{\sqrt{3}}{2}$

94. IV. Trigonometric Graphs

8. If $\sin^{-1}(\frac{5}{8}) = A$, then

*(1) $\sin A = \frac{5}{8}$ (2) $\sin A = \frac{8}{5}$ (3) $\cos A = \frac{5}{8}$ (4) $\cos A = \frac{8}{5}$

$y = \sin^{-1} x$ is equivalent to $x = \sin y$.
(Same as: $\log_2 8 = 3$ is equivalent to $2^3 = 8$)

---

9. In physics class, Eva noticed the pattern shown in the accompanying diagram on an oscilloscope.

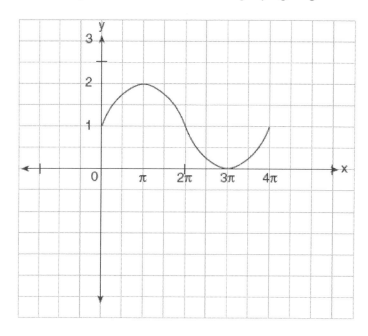

Which equation best represents the pattern shown on this oscilloscope?

*(1) $y = \sin(\frac{1}{2} x) + 1$ (3) $y = 2 \sin x + 1$

(2) $y = \sin x + 1$ (4) $y = 2 \sin(-\frac{1}{2} x) + 1$

---
1. Maximum Value $y = 2$, Minimum Value $y = 0$

   Amplitude: $a = \dfrac{\text{Max} - \text{Min}}{2} = \dfrac{2 - 0}{2} = 1$

2. Period $P = 4\pi$, Frequency: $b = \dfrac{2\pi}{P} = \dfrac{2\pi}{4\pi} = \dfrac{1}{2}$

3. Vertical shift : $d = 1$

$$y = a\sin bx + d = \sin(\frac{1}{2} x) + 1$$

Answers V. Trigonometric Applications

1. The expression $\cos 4x \cos 3x + \sin 4x \sin 3x$ is equivalent to
   (1) $\sin x$ (2) $\sin 7x$ *(3) $\cos x$ (4) $\cos 7x$

   $\cos 4x \cos 3x + \sin 4x \sin 3x = \cos(4x - 3x) = \cos x$

2. The expression $\cos^2 \theta - \cos 2\theta$ is equivalent to
   *(1) $\sin^2 \theta$ (2) $-\sin^2 \theta$ (3) $\cos^2 \theta + 1$ (4) $-\cos^2 \theta - 1$

   $\cos^2 \theta - \cos 2\theta = \cos^2 \theta - (\cos^2 \theta - \sin^2 \theta) = \sin^2 \theta$

3. If $\sin A = \dfrac{2}{3}$ where $0° < A < 90°$, what is the value of $\sin 2A$?

   (1) $\dfrac{2\sqrt{5}}{3}$ (2) $\dfrac{2\sqrt{5}}{9}$ *(3) $\dfrac{4\sqrt{5}}{9}$ (4) $-\dfrac{4\sqrt{5}}{9}$

   $\sin^2 A + \cos^2 A = 1$

   $\cos A = \sqrt{1 - \sin^2 A} = \sqrt{1 - \dfrac{4}{9}} = \sqrt{\dfrac{9-4}{3}} = \dfrac{\sqrt{5}}{3}$ positive since $\angle A$ is in Quadrant I

   $\sin 2A = 2\sin A \cdot \cos A = 2 \cdot \dfrac{2}{3} \cdot \dfrac{\sqrt{5}}{3} = \dfrac{4\sqrt{5}}{9}$

4. What are the values of θ in the interval $0° \leq \theta < 360°$ that satisfy the equation $\tan \theta - \sqrt{3} = 0$?
   *(1) 60°, 240° (3) 72°, 108°, 252°, 288°
   (2) 72°, 252° (4) 60°, 120°, 240°, 300°

   $\tan \theta = \sqrt{3}$
   θ is in the Quadrant I and III.
   $\theta = 60°$ and $\theta = 180° + 60° = 240°$

5. In $\triangle ABC$, $m\angle A = 120$, $b = 10$, and $c = 18$. What is the area of $\triangle ABC$ to the *nearest square inch*?
   (1) 52 *(2) 78 (3) 90 (4) 156

   Area $= \dfrac{1}{2} bc \cdot \sin A = \dfrac{1}{2} \cdot 10 \cdot 18 \cdot \sin 120° \approx 78$

V. Trigonometric Applications

6. In $\triangle ABC$, $a = 3$, $b = 5$, and $c = 7$. What is $m\angle C$?
(1)   22   (2)   38   (3)   60   *(4)   120

$$\cos C = \frac{a^2 + b^2 - c^2}{2ab} = \frac{3^2 + 5^2 - 7^2}{2 \cdot 3 \cdot 5} = -0.5 \quad (\angle C \text{ is in Quadrant II})$$
$$m\angle C = \cos^{-1}(-0.5) = 120$$

7. The sides of a parallelogram measure 10 cm and 18 cm. One angle of the parallelogram measures 46 degrees. What is the area of the parallelogram, to the *nearest square centimeter*?
(1)   65   (2)   125   *(3)   129   (4)   162

A parallelogram consists of two congruent triangles.
$$\text{Area} = 2 \cdot \frac{1}{2} ab \cdot \sin C = 10 \cdot 18 \cdot \sin 46° \approx 129$$

8. In $\triangle ABC$, $m\angle A = 74$, $a = 59.2$, and $c = 60.3$. What are the two possible values for $m\angle C$, to the *nearest tenth*?
(1)   73.7 and 106.3   (2)   73.7 and 163.7   *(3)   78.3 and 101.7   (4)   78.3 and 168.3

$$\frac{\sin C}{c} = \frac{\sin A}{a}$$
$$\sin C = \frac{c \cdot \sin A}{a} = \frac{60.3 \cdot \sin 74°}{59.2} = 0.9791229775$$
$$m\angle C = \sin^{-1}(0.9791229775)$$
$$m\angle C = 78.3 \quad \text{or} \quad m\angle C = 180 - 78.3 = 101.7$$

9. How many distinct triangles can be formed if $m\angle A = 35$, $a = 10$, and $b = 13$?
(1)   1   *(2)   2   (3)   3   (4)   0

$$\frac{\sin B}{b} = \frac{\sin A}{a}$$
$$\sin B = \frac{13 \sin 35°}{10}$$
$$m\angle B \approx 48.2 \quad \text{or} \quad m\angle B \approx 180 - 48.2 = 131.8$$
Check $48.2 + 35 < 180$
         $131.8 + 35 < 180$
Two distinct triangles can be formed.

Answers   V. Trigonometric Applications

Show Work:

1. If $\tan A = \dfrac{2}{3}$ and $\sin B = \dfrac{5}{\sqrt{41}}$ and angles A and B are in Quadrant I, find the value of $\tan(A+B)$.

$$\cos B = \sqrt{1 - \sin^2 B} = \sqrt{1 - \dfrac{25}{41}} = \dfrac{4}{\sqrt{41}}$$

$$\tan B = \dfrac{\sin B}{\cos B} = \dfrac{\frac{5}{\sqrt{41}}}{\frac{4}{\sqrt{41}}} = \dfrac{5}{4}$$

$$\tan(A+B) = \dfrac{\tan A + \tan B}{1 - \tan A \cdot \tan B} = \dfrac{\frac{2}{3} + \frac{5}{4}}{1 - \frac{2}{3} \cdot \frac{5}{4}} = \dfrac{\frac{8+15}{12}}{\frac{12-10}{12}} = \dfrac{23}{12} \cdot \dfrac{12}{2} = \dfrac{23}{2}$$

2. Starting with $\sin^2 A + \cos^2 A = 1$, derive the formula $\tan^2 A + 1 = \sec^2 A$.

$$\dfrac{\sin^2 A + \cos^2 A}{\cos^2 A} = \dfrac{1}{\cos^2 A}$$

$$\tan^2 A + 1 = \sec^2 A$$

3. Find all values of θ in the interval $0° \leq \theta < 360°$ that satisfy the equation $\sin 2\theta = \sin \theta$.

$$2\sin\theta\cos\theta = \sin\theta$$
$$2\sin\theta\cos\theta - \sin\theta = 0$$
$$\sin\theta(2\cos\theta - 1) = 0$$

$\sin\theta = 0$ | $2\cos\theta - 1 = 0$
$\theta = 0°, \theta = 180°$ | $\cos\theta = \dfrac{1}{2}$
                $\theta = 60°, \theta = 360° - 60° = 300°$

$\{0°, 60°, 180°, 300°\}$

4. Solve the equation $2\tan C - 3 = 3\tan C - 4$ algebraically for all values of C in the interval $0° \leq C < 360°$.

$2\tan C - 3 = 3\tan C - 4$
$\tan C = 1$
$C = 45°$ and $C = 180° + 45° = 225°$
$\{45°, 225°\}$

5. Two sides of a parallelogram are 24 feet and 30 feet. The measure of the angle between these sides is 57°. Find the area of the parallelogram, to the *nearest square foot*.

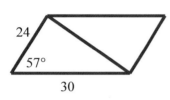

Area of a Triangle:
$$K = \frac{1}{2}ab \cdot \sin C$$

Area of the parallelogram:
$A = 2K = ab \cdot \sin C = 24 \cdot 30 \cdot \sin 57° \approx \mathbf{604}$

6. In a triangle, two sides that measure 6 cm and 10 cm form an angle that measures 80°. Find, to the *nearest degree*, the measure of the smallest angle in the triangle.

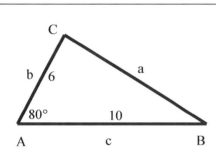

$a^2 = b^2 + c^2 - 2bc \cdot \cos A$
$= 6^2 + 10^2 - 2 \cdot 6 \cdot 10 \cdot \cos 80°$
$= 115.1622187$
$a = 10.73136611$

Compare, $\angle B$ is the smallest angle.
$$\frac{\sin B}{b} = \frac{\sin A}{a}$$
$$\sin B = \frac{b \cdot \sin A}{a} = \frac{6\sin 80°}{10.73136611} \approx 0.5506$$
$B = \sin^{-1}(0.5506) \approx \mathbf{33°}$

7. Two forces of 25 newtons and 85 newtons acting on a body form an angle of 55°. Find the magnitude of the resultant force, to the *nearest hundredth of a newton*. Find the measure, to the *nearest degree*, of the angle formed between the resultant and the larger force.

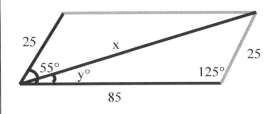

Use parallelogram to solve the force problem.

(1) Law of Cosines:
$$a^2 = b^2 + c^2 - 2bc \cdot \cos A$$
$$x^2 = 85^2 + 25^2 - 2 \cdot 85 \cdot 25 \cdot \cos 125°$$
x ≈ 101.43

(2) Law of Sines:
$$\frac{\sin y°}{25} = \frac{\sin 125°}{101.43}$$
$$\sin y° = \frac{25 \cdot \sin 125°}{101.43} \approx 0.2019$$
$$y° = \sin^{-1}(0.2019) \approx \mathbf{12°}$$

8. In $\triangle ABC$, $m\angle A = 32$, $a = 12$, and $b = 10$. Find the measures of the missing angles and side of $\triangle ABC$. Round each measure to the *nearest tenth*.

$$\frac{\sin B}{b} = \frac{\sin A}{a}$$
$$\sin B = \frac{10 \sin 32°}{12}$$
$m\angle B \approx 26.2$ or $m\angle B \approx 180 - 26.2 = 153.8$
Check $26.2 + 32 < 180$ OK
          $153.8 + 32 > 180$ Rejected
m∠B = 26.2
m∠C = 180 - 32 - 26.2 = 121.8
$$\frac{c}{\sin C} = \frac{a}{\sin A}$$
$$\frac{c}{\sin 121.8°} = \frac{12}{\sin 32°}$$
c ≈ 19.2

VI. Probability

1. A dartboard is shown in the diagram below. The two lines intersect at the center of the circle, and the central angle in sector 2 measures $\frac{2\pi}{3}$.

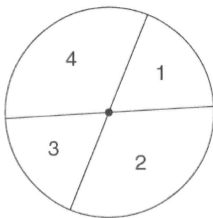

If darts thrown at this board are equally likely to land anywhere on the board, what is the probability that a dart that hits the board will land in either sector 1 or sector 3?

(1) $\frac{1}{6}$ *(2) $\frac{1}{3}$ (3) $\frac{1}{2}$ (4) $\frac{2}{3}$

> The central angle in sector 1 is $\pi - \frac{2\pi}{3} = \frac{\pi}{3}$.
>
> The central angle in sector 3 is also equal to $\frac{\pi}{3}$.
>
> $P(1 \text{ or } 3) = \dfrac{\frac{\pi}{3} + \frac{\pi}{3}}{2\pi} = \dfrac{1}{3}$

2. A four-digit serial number is to be created from the digits 0 through 9.
How many of these serial numbers can be created if 0 can *not* be the first digit, no digit may be repeated, and the last digit must be 5?

*(1) 448   (2) 504   (3) 2,240   (4) 2,520

> $8 \bullet 8 \bullet 7 \bullet 1 = \mathbf{448}$
> The last digit has 1 choice.
> Then first digit has 8 choices (excluding 5 and 0).
> Then the second digit has 8 choices left and the third digit has 7 choices left.

3. Which formula can be used to determine the total number of different eight-letter arrangements that can be formed using the letters in the word *DEADLINE*?

(1) $8!$ (2) $\dfrac{8!}{4!}$ (3) $\dfrac{8!}{2!+2!}$ *(4) $\dfrac{8!}{2! \cdot 2!}$

> 8! for eight-letter arrangements, divided by 2!•2! for 2 identical E's and 2 identical D's

Answers   VI. Probability

4. Twenty different cameras will be assigned to several boxes. Three cameras will be randomly selected and assigned to box A. Which expression can be used to calculate the number of ways that three cameras can be assigned to box A?

(1) $20!$ (2) $\dfrac{20!}{3!}$ *(3) $_{20}C_3$ (4) $_{20}P_3$

The order of the three different cameras is not a concern in this question.

---

5. The principal would like to assemble a committee of 8 students from the 15-member student council. How many different committees can be chosen?

(1)   120   *(2)   6,435   (3)   32,432,400   (4)   259,459,200

$$_{15}C_8 = \dfrac{15 \cdot 14 \cdot 13 \cdot 12 \cdot 11 \cdot 10 \cdot 9 \cdot 8}{8!} = 6,435 \quad \text{or}$$

Using Graphing Calculator: [15] [MATH] PRE / 3: $_nC_r$ [8] [ENTER]

---

6. Three marbles are to be drawn at random, without replacement, from a bag containing 15 red marbles, 10 blue marbles, and 5 white marbles. Which expression can be used to calculate the probability of drawing 2 red marbles and 1 white marble from the bag?

*(1) $\dfrac{_{15}C_2 \bullet {_5C_1}}{_{30}C_3}$ (2) $\dfrac{_{15}P_2 \bullet {_5P_1}}{_{30}C_3}$ (3) $\dfrac{_{15}C_2 \bullet {_5C_1}}{_{30}P_3}$ (4) $\dfrac{_{15}P_2 \bullet {_5P_1}}{_{30}P_3}$

The number of the sample space: $n(S) = {_{30}C_3}$,   The number of the event: $n(E) = {_{15}C_2} \bullet {_5C_1}$

$$P(E) = \dfrac{n(E)}{n(S)} = \dfrac{_{15}C_2 \bullet {_5C_1}}{_{30}C_3}$$

---

7. What is the fourth term in the expansion of $(3x - 2)^5$?

*(1) $-720x^2$ (2) $-240x$ (3) $720x^2$ (4) $1,080x^3$

The rth term: $_nC_{(r-1)} X^{n-(r-1)} Y^{(r-1)}$ here $(r-1) = 4 - 1 = 3$, $n = 5$

$_5C_3 \bullet X^{(5-3)} Y^3 = {_5C_3} \bullet X^2 Y^3 = {_5C_3} \bullet (3x)^2 (-2)^3$ here $X = 3x$, $Y = -2$

$$= \dfrac{5 \bullet 4 \bullet 3}{3 \bullet 2 \bullet 1} (9x^2)(-8) = -720x^2$$

VI. Probability

Show Work:

1. The letters of any word can be rearranged. Carol believes that the number of different 9-letter arrangements of the word "TENNESSEE" is greater than the number of different 7-letter arrangements of the word "VERMONT." Is she correct? Justify your answer.

> No, Carol is wrong.
> The number of arrangements of "TENNESSEE": $\dfrac{9!}{4! \cdot 2! \cdot 2!} = 3{,}780$
> The number of arrangements of "VERMONT": $7! = 5{,}040$
> (Hint: using graphing calculator for factorial 7!: [7] [MATH] PRB / 4: ! [ENTER])

2. Find the total number of different twelve-letter arrangements that can be formed using the letters in the word *PENNSYLVANIA*.

> 12! for twelve-letter arrangements, divided by 3!•2! for 3 identical N's and 2 identical A's
> $\dfrac{12!}{3! \cdot 2!} = 39{,}916{,}800$
> (using graphing calculator for 12!: [12] [MATH] PRB / 4: ! [ENTER])

3. A committee of 5 members is to be randomly selected from a group of 9 teachers and 20 students. Determine how many different committees can be formed if 2 members must be teachers and 3 members must be students.

> $_9C_2 \cdot {_{20}C_3} = \dfrac{9 \cdot 8}{2 \cdot 1} \cdot \dfrac{20 \cdot 19 \cdot 18}{3 \cdot 2 \cdot 1} = \mathbf{41{,}040}$

4. The members of a men's club have a choice of wearing black or red vests to their club meetings. A study done over a period of many years determined that the percentage of black vests worn is 60%. If there are 10 men at a club meeting on a given night, what is the probability, to the *nearest thousandth*, that *at least* 8 of the vests worn will be black?

> P(Black) = 0.6, P(Not Black) = 0.4
> The probability of 8 black vests: $P(8) = {_{10}C_8}(0.6)^8 \cdot (0.4)^2$
> The probability of 9 black vests: $P(9) = {_{10}C_9}(0.6)^9 \cdot (0.4)^1$
> The probability of 10 black vests: $P(10) = {_{10}C_{10}}(0.6)^{10} \cdot (0.4)^0$
> P(at least 8) = P(8) + P(9) + P(10)
> $= \dfrac{10 \cdot 9}{2 \cdot 1} \cdot (0.6)^8 \cdot (0.4)^2 + \dfrac{10}{1} \cdot (0.6)^9 \cdot (0.4)^1 + 1 \cdot (0.6)^{10} \cdot (0.4)^0$
> $\approx \mathbf{0.167}$
> (Hint: $_{10}C_8 = {_{10}C_2}$, $_{10}C_9 = {_{10}C_1}$, $_{10}C_{10} = 1$)

Answers   VI. Probability

5. The probability that the Stormville Sluggers will win a baseball game is $\frac{2}{3}$. Determine the probability, to the *nearest thousandth*, that the Stormville Sluggers will win *at least* 6 of their next 8 games.

$p = \frac{2}{3}$, $q = \frac{1}{3}$, $n = 8$

$P(6) = {}_8C_6 \left(\frac{2}{3}\right)^6 \left(\frac{1}{3}\right)^2 = 28 \cdot \left(\frac{2^6}{3^8}\right)$

$P(7) = {}_8C_7 \left(\frac{2}{3}\right)^7 \left(\frac{1}{3}\right)^1 = 8 \cdot \left(\frac{2^7}{3^8}\right)$

$P(8) = {}_8C_8 \left(\frac{2}{3}\right)^8 \left(\frac{1}{3}\right)^0 = 1 \cdot \left(\frac{2^8}{3^8}\right)$

$P(\text{at least } 6) = P(6) + P(7) + P(8)$

$= \frac{28 \cdot 2^6 + 8 \cdot 2^7 + 2^8}{3^8}$

$= \frac{3072}{6561} \approx \mathbf{0.468}$

6. A study shows that 35% of the fish caught in a local lake had high levels of mercury. Suppose that 10 fish were caught from this lake. Find, to the *nearest tenth of a percent*, the probability that *at least* 8 of the 10 fish caught did *not* contain high levels of mercury.

E: Event of *not* high levels of mercury
~E: Event of high levels of mercury
$P(E) = 0.65$, $P(\sim E) = 0.35$
$P(8) = {}_{10}C_8 (0.65)^8 \cdot (0.35)^2$
$P(9) = {}_{10}C_9 (0.65)^9 \cdot (0.35)^1$
$P(10) = {}_{10}C_{10} (0.66)^{10} \cdot (0.35)^0$
$P(\text{at least } 8) = P(8) + P(9) + P(10) \approx 0.262 \approx \mathbf{26.2\%}$
(Hint: ${}_{10}C_8 = {}_{10}C_2 = \frac{10 \cdot 9}{2 \cdot 1} = 45$, ${}_{10}C_9 = {}_{10}C_1 = 10$, ${}_{10}C_{10} = 1$)

7. Write the binomial expansion of $(2x - 1)^5$ as a polynomial in simplest form.

$(a + b)^5$
$= {}_5C_0 a^5 b^0 + {}_5C_1 a^4 b^1 + {}_5C_2 a^3 b^2 + {}_5C_3 a^2 b^3 + {}_5C_4 a^1 b^4 + {}_5C_5 a^0 b^5$
$= a^5 b^0 + 5a^4 b^1 + 10a^3 b^2 + 10a^2 b^3 + 5a^1 b^4 + a^0 b^5$
$= (2x)^5(-1)^0 + 5(2x)^4(-1)^1 + 10(2x)^3(-1)^2 + 10(2x)^2(-1)^3 + 5(2x)^1(-1)^4 + (2x)^0(-1)^5$
$= \mathbf{32x^5 - 80x^4 + 80x^3 - 40x^2 + 10x - 1}$

104.   VII. Statistics

1. A survey completed at a large university asked 2,000 students to estimate the average number of hours they spend studying each week. Every tenth student entering the library was surveyed. The data showed that the mean number of hours that students spend studying was 15.7 per week. Which characteristic of the survey could create a bias in the results?
(1) the size of the sample
(2) the size of the population
(3) the method of analyzing the data
*(4) the method of choosing the students who were surveyed

> Choosing the students entering the library creates a bias.
> Choosing the students at the entrance to the school will be fair.

2. Which task is *not* a component of an observational study?
   (1) The researcher decides who will make up the sample.
   (2) The researcher analyzes the data received from the sample.
   (3) The researcher gathers data from the sample, using surveys or taking measurements.
*(4) The researcher divides the sample into two groups, with one group acting as a control group.

> The difference between "Observation" and "Controlled Experiment" is that observation does not use a controlled sample as a benchmark.

3. The table below shows the first-quarter averages for Mr. Harper's statistics class.

Statistics Class Averages

Quarter Averages	Frequency
99	1
97	5
95	4
92	4
90	7
87	2
84	6
81	2
75	1
70	2
65	1

What is the population variance for this set of data?
(1) 8.2   (2) 8.3   *(3) 67.3   (4) 69.3

> using graphing calculator to find the standard deviation $\delta_x \approx 8.2$
> population variance: $v = (\delta_x)^2 \approx \mathbf{67.3}$
> (Hint: refer to page 21)

4. The lengths of 100 pipes have a normal distribution with a mean of 102.4 inches and a standard deviation of 0.2 inch. If one of the pipes measures exactly 102.1 inches, its length lies
*(1) below the 16th percentile
(2) between the 50th and 84th percentiles
(3) between the 16th and 50th percentiles
(4) above the 84th percentile

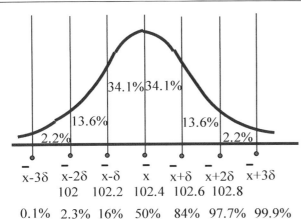

mean $\bar{x}$ = 102.4 is at the 50th percentile;
1 standard deviation below the mean $\bar{x} - \delta$
= 102.4 - 0.2 = 102.2 is at the 16th percentile;
102.1 is less than 102.2, therefore is below the 16th percentile.

5. An amateur bowler calculated his bowling average for the season. If the data are normally distributed, about how many of his 50 games were within one standard deviation of the mean?
(1)   14   (2)   17   *(3)   34   (4)   48

For a normal distribution, approximately 68% of the data values lie within one standard deviation from the mean.
$$50 \cdot 68\% = 34$$

6. Which value of r represents data with a strong negative linear correlation between two variables?
(1)   −1.07   *(2)   −0.89   (3)   −0.14   (4)   0.92

The range of the correlation coefficient r is between -1 and 1.

106.   **VII. Statistics**

Show Work:

1. Howard collected fish eggs from a pond behind his house so he could determine whether sunlight had an effect on how many of the eggs hatched. After he collected the eggs, he divided them into two tanks. He put both tanks outside near the pond, and he covered one of the tanks with a box to block out all sunlight. State whether Howard's investigation was an example of a controlled experiment, an observation, or a survey. Justify your response.

> The investigation was an example of a controlled experiment.
> Controlled experiments consist of two groups of data, one of them is served as a benchmark.

2. Assume that the ages of first-year college students are normally distributed with a mean of 19 years and standard deviation of 1 year.

(1) To the *nearest integer*, find the percentage of first-year college students who are between the ages of 18 years and 20 years, inclusive.

(2) To the *nearest integer*, find the percentage of first-year college students who are 20 years old or older.

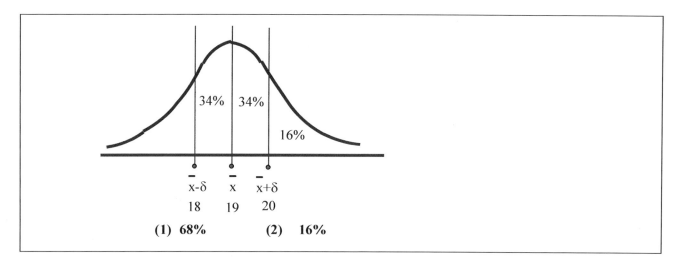

(1) 68%   (2) 16%

3. The scores of one class on the Unit 2 mathematics test are shown in the table below.

Unit 2 Mathematics Test

Test Score	Frequency
96	1
92	2
84	5
80	3
76	6
72	3
68	2

Find the population standard deviation of these scores, to the *nearest tenth*.

Using Graphing Calculator:
Enter Test Score in L_1 ; Enter Frequency in L_2.
[ENTER] CALC / 1: 1-Var [ENTER] L_1 , L_2 [ENTER]

$$\delta_x \approx 7.4$$

(*This table represents univariate data - one variate.
 Frequency is the number of times that a value occurs, it does not represent another variate.)

---

4. The table below shows the results of an experiment involving the growth of bacteria.

Time (x) (in minutes)	1	3	5	7	9	11
Number of Bacteria (y)	2	25	81	175	310	497

Write a power regression equation for this set of data, rounding all values to *three decimal places*. Using this equation, predict the bacteria's growth, to the *nearest integer*, after 15 minutes.

Using Graphing Calculator:
(1) Enter the data into L_1 and L_2.
(2) Use the Power Regression model:
   [STAT] CALC / A: PwrReg [ENTER] L_1 , L_2 [ENTER]

$$y = ax^b$$
   a = 2.000877106, b = 2.298056258
   rounded to three decimal places
   $$y = 2.001 \bullet x^{2.298}$$
(3) Use this equation for x = 15
   $$y = 2.001 \bullet (15)^{2.298} \approx \mathbf{1,009}$$
(*Scatter plot and graph are not required in this question.)

5. The table below shows the number of new stores in a coffee shop chain that opened during the years 1986 through 1994.

Year	Number of New Stores
1986	14
1987	27
1988	48
1989	80
1990	110
1991	153
1992	261
1993	403
1994	681

Using $x = 1$ to represent the year 1986 and y to represent the number of new stores, write the exponential regression equation for these data. Round all values to the *nearest thousandth*.

Using Graphing Calculator:
Enter the data into L_1 and L_2.

Use the Exponential Regression model:
 [STAT] CALC / 0: ExpReg [ENTER] L_1 , L_2 [ENTER]

$$y = 10.596 \cdot 1.586^x$$

Also available:

Student's Choice
Regents Review Integrated Algebra   ISBN: 9781453880982

Student's Choice
Regents Review Geometry   ISBN: 9781453709993

Teacher's Choice
Math Regents Review   ISBN: 9781450562843
(Handy reference book for math teachers and college students)

Made in the USA
Lexington, KY
19 March 2012